Algebraic Number Theory

Algebraic Number Theory

Editor

Aneta Hajek

Algebraic Number Theory

Edited by **Aneta Hajek**

Printed in 2017

ISBN: 978-1-68117-183-8

Library of Congress Control Number: 2015949110

© 2016 by

SCITUS Academics LLC,

616, Corporate Way, Suite 2, 4766,

Valley Cottage, NY 10989

www.scitusacademics.com

Preface

Algebraic number theory is the branch of number theory that deals with algebraic numbers. Historically, algebraic number theory developed as a set of tools for solving problems in elementary number theory, namely Diophantine equations (i.e., equations whose solutions are integers or rational numbers). More recently, algebraic number theory has developed into the abstract study of algebraic numbers and number fields themselves, as well as their properties. Algebraic number theory is a major branch of number theory that studies algebraic structures related to algebraic integers. This is generally accomplished by considering a ring of algebraic integers O in an algebraic number field K/Q, and studying their algebraic properties such as factorization, the behaviour of ideals, and field extensions. In this setting, the familiar features of the integers—such as unique factorization—need not hold. The virtue of the primary machinery employed—Galois theory, group cohomology, group representations, and L-functions—is that it allows one to deal with new phenomena and yet partially recover the behaviour of the usual integers.

The higher reaches of algebraic number theory are now one of the crown jewels of mathematics. But algebraic number theory is not merely interesting in itself. It has become an important tool over a wide range of pure mathematics, and many of ideas involved generalize, for example to algebraic geometry. This book is intended both for number theorist and more generally for working algebraists.

Contents

Two-Primary Algebraic K-Theory of Pointed Spaces 1

Probability Theory Predicts that Chunking into groups of three
or four items Increases the Short-term Memory Capacity 91

Applications to Cryptography of Twisting Commutative
Algebraic Groups 111

Image Mathematics—Mathematical Intervening
Principle Based on "Yin Yang Wu Xing" Theory in
Traditional Chinese Mathematics (I) 151

First Review of Articles on Rhotrix Theory since Its Inception 209

Monty Hall Problem and the Principle of Equal Probability in
Measurement Theory 225

Contractions of Certain Lie Algebras in the
Context of the DLF-Theory 245

Index 267

Two-Primary Algebraic K-Theory of Pointed Spaces

John Rognes
Department of Mathematics, University of Oslo,
Blindern, N-0316 Oslo, Norway

ABSTRACT

We compute the mod 2 cohomology of Waldhausen's algebraic K-theory spectrum $A(*)$ of the category of finite pointed spaces, as a module over the Steenrod algebra. This also computes the mod 2 cohomology of the smooth Whitehead spectrum of a point, denoted $Wh^{Diff}(*)$. Using an Adams spectral sequence we compute the 2-primary homotopy groups of these spectra in dimensions $* \leq 18$, and up to extensions in dimensions $19 \leq * \leq 21$. As applications we show that the linearization map $L : A(*) \to K(\mathbb{Z})$ induces the zero homomorphism in mod 2 spectrum cohomology in positive dimensions, the space level Hatcher–Waldhausen map $hw : G/O \to \Omega Wh^{Diff}(*)$ does not admit a four-fold delooping, and there is a 2-complete spectrum map $M : Wh^{Diff}(*) \to \sum g/o_{\oplus}$ which is precisely 9-connected. Here $g/o_{\oplus}$ is a spectrum whose underlying space has the 2-complete homotopy type of G/O.

INTRODUCTION

Let $A(X)$ be Waldhausen's algebraic K-theory of spaces functor evaluated on the space X, see [53]. When X is a manifold, $A(X)$ provides the fundamental link between algebraic K-theory and the geometric topology of X—in particular with the concordance space, the h-cobordism space and the automorphism space of X, see [55]. We are therefore interested in evaluating its homotopy type. It is the aim of this paper to compute the 2-primary homotopy type of $A(X)$ in the case when $X=*$ is the one-point space. We achieve this by computing the mod 2 spectrum cohomology of $A(*)$ as a module over the mod 2 Steenrod algebra. The result is a complete calculation valid in all dimensions; we also compute the homotopy groups of $A(*)$ modulo odd torsion in dimensions $* \leq 18$, and up to extensions in dimensions $19 \leq * \leq 21$.

We begin by discussing some definitions and interpretations of $A(X)$, in order to explain why this is an important homotopy type.

One way to define $A(X)$ is as the algebraic K-theory of a category with cofibrations and weak equivalences $R_f(X)$, whose objects are retractive spaces over X subject to a relative finiteness condition, see [57]. When $X=*$ this category $R_f(*)$ is the category of finite pointed CW-complexes and pointed cellular maps, and is the category of pointed spaces alluded to in the title. The cofibrations are the cellular embeddings, and the weak equivalences are the homotopy equivalences.

Let $hR_f(X)$ be the subcategory of $R_f(X)$ obtained by restricting the morphisms to be homotopy equivalences, and let $|hR_f(X)|$ denote its geometric realization. As a space, $A(X)$ is defined as the loop space $\Omega |hS_. R_f(X)|$, where $S_.$ is Waldhausen's simplicial construction of the same name. This construction can be iterated, and in fact $A(X)$ is an infinite loop space with nth delooping $|hS_.^{(n)} R_f(X)|$ for each $n \geq 1$. There is a canonical map

$$e: |hR_f(X)| \to A(X)$$

Two-Primary Algebraic K-Theory of Pointed Spaces

From the geometric realization of the category of finite retractive spaces over X and homotopy equivalences to the infinite loop space A(X).

There is a natural isomorphism $\pi_0 A(X) \cong \mathbb{Z}$, and for every object $Y \in hR_f(X)$ the image under $\pi_0(e)$ of the corresponding point in $|hR_f(X)|$ is the Euler characteristic $x(Y, X) = \tilde{x}(Y/X)$ of Y relative to X. From this point of view the map e is a lift of the usual Euler characteristic that takes values in the integers, to a map which takes values in the infinite loop space A(X). Furthermore, a diagram of spaces and homotopy equivalences given as a functor $F: I \to hR_f(X)$ gives rise to a map $e|F|: |\lambda| \to A(*)$, which will detect more information than just the Euler characteristics of the individual spaces in the diagram. For example, a finite pointed G-CW complex Y gives rise to a map BG→A(*) whose homotopy class is a refined invariant of Y. We think of e as a homotopy theoretic improvement on the Euler characteristic, able also to detect information about diagrams of spaces and homotopy equivalences, rather than just individual spaces, and A(X) is the receptacle for this improved Euler characteristic.

In fact, A(*) is a kind of universal receptacle for homotopy invariants of finite pointed spaces that take values in infinite loop spaces and are subject to the following additivity condition: for each cofiber sequence Y'→Y→Y'' we have [Y']+[Y'']=[Y] where $[Y] \in \pi_0 A(*)$ denotes the path component in A(*) of the invariant applied to Y. Of course, the corresponding universal invariant taking values in an abelian group is just the reduced Euler characteristic. We shall not make the universality claim more precise in this introduction, but note that a similar discussion applies for A(X) and suitably additive homotopy invariants of finite retractive spaces over X.

Hereafter, it will be more convenient to work with spectra than with infinite loop spaces. The given deloopings of the infinite loop space A(X) determine a unique connective spectrum, and from now on A(X) will refer to this spectrum. The body of this paper is also written in terms of spectra rather than infinite loop spaces, partly because a few non-connective spectra will appear.

Suspension of retractive spaces over X induces an equivalence on the level of algebraic K-theory, and soA(X) can also be considered as the algebraic K-theory of a category of spectra over X. It is simplest to make this precise for X=*, when A(*) is equivalent to the algebraic K-theory of the category of finite CW-spectra, with respect to suitable notions of cofibrations and stable equivalences, see [56].

Let $\mathbb{S}$ be the sphere spectrum in some good closed symmetric monoidal category of spectra and spectrum maps, for example the S-modules of [19] or the Γ-spaces of [27] and [47]. In either case the ring spectrum $\mathbb{S}$ is a monoid object with respect to the internal smash product, and a spectrum is a module over $\mathbb{S}$, so we can sensibly refer to spectra as $\mathbb{S}$-modules. Then A(*) can be described as the algebraic K-theory of a category of $\mathbb{S}$-modules subject to suitable finiteness conditions, and briefly A(*) is the algebraic K-theory of the ring spectrum $\mathbb{S}$. See [8] for a discussion in terms of FSPs.

More generally, for a unital and associative ring spectrum A we may consider a category of finitely generated free A-modules and form its algebraic K-theory, see [15]. These ring spectra are the monoids in one of the symmetric monoidal categories of spectra considered above, and may conveniently be called $\mathbb{S}$-algebras. For each ring R in the algebraic sense, the Eilenberg–Mac Lane spectrum HR is an $\mathbb{S}$-algebra whose algebraic K-theory agrees with Quillen's K(R), see [39]. For a simplicial monoid G the unreduced suspension spectrum $\Sigma^{\infty}(G_{+})$ is an $\mathbb{S}$-algebra whose algebraic K-theory agrees with Waldhausen's A(X) forX=BG. Thus $\mathbb{S}$-algebras encompass the previous examples of inputs for algebraic K-theory. Now $\mathbb{S}$ is a commutative $\mathbb{S}$-algebra, so its algebraic K-theory K ($\mathbb{S}$)=A(*) is itself a ring spectrum, and furthermore the algebraic K-theory K(A) of any $\mathbb{S}$-algebra is a module spectrum over A(*). Hence, every algebraic K-theory spectrum considered so far is a module spectrum over A(*), which further emphasizes the special role played by A(*).

Two-Primary Algebraic K-Theory of Pointed Spaces

The relationship of $A(X)$ to geometric topology is based on the splitting of spectra $A(X) \simeq \Sigma^\infty(X_+) \vee \mathrm{Wh}^{\mathrm{Diff}}(X)$ for the smooth category, and the cofiber sequence of spectra

$$A(*) \wedge X_+ \xrightarrow{\alpha} A(X) \to \mathrm{Wh}^{\mathrm{PL}}(X)$$

for the piecewise linear category, see [55] and [58]. Here α is the assembly map, one construction of which uses that $A(X)$ is a homotopy functor in X, see [62].

The spectra $\mathrm{Wh}^{\mathrm{Diff}}(X)$ and $\mathrm{Wh}^{\mathrm{PL}}(X)$ are the smooth and PL Whitehead spectra, respectively. The topological Whitehead spectrum $\mathrm{Wh}^{\mathrm{Top}}(X)$ is equivalent to the PL one by Kirby and Siebenmann [24] and Burghelea and Lashof [13]. Thus, knowledge of $A(*)$ determines $\mathrm{Wh}^{\mathrm{Diff}}(*)$ and is the ingredient needed to pass from $A(X)$ to $\mathrm{Wh}^{\mathrm{PL}}(X) \simeq \mathrm{Wh}^{\mathrm{Top}}(X)$. The underlying infinite loop spaces of these Whitehead spectra are called Whitehead spaces, and it is perhaps more common to work in terms of these.

When X is a smooth manifold, $\Omega^\infty \mathrm{Wh}^{\mathrm{Diff}}(X)$ gives the homotopy functor that best approximates the space $C^{\mathrm{Diff}}(X)$ of smooth concordances (= pseudoisotopies) of X. By Igusa's stability theorem [23] there is a stabilization map

$$\Sigma_X^{\mathrm{Diff}} : C^{\mathrm{Diff}}(X) \to \Omega^2 \Omega^\infty \mathrm{Wh}^{\mathrm{Diff}}(X)$$

Which is at least roughly n/3-connected where n is the dimension of X. Similar results relate $\mathrm{Wh}^{\mathrm{PL}}(X)$ and $\mathrm{Wh}^{\mathrm{Top}}(X)$ to the PL- and topological concordance spaces $C^{\mathrm{PL}}(X)$ and $C^{\mathrm{Top}}(X)$ when X is a PL- or topological manifold, respectively.

Furthermore, there is a geometrically significant involution on $A(X)$, related through the Whitehead spectra to the involution on concordance spaces arising from 'turning a concordance upside-down', see [21] and [52]. By Weiss and Williams [61] there is a map

$$\Phi_X^{\mathrm{Diff}} : \widetilde{\mathrm{Diff}}(X)/\mathrm{Diff}(X) \to \Omega^\infty(EC_{2+} \wedge_{C_2} \Omega\mathrm{Wh}^{\mathrm{Diff}}(X))$$

Which is at least as connected as the stabilization map considered by Igusa. The C_2-action on $\Omega\mathrm{Wh}^{\mathrm{Diff}}(X)$ on the right is given by the involution, and the homotopy orbit construction is formed on the spectrum level. This is a space level interpretation of the output of the Hatcher spectral sequence [21], which works on the level of homotopy groups.

The space $\widetilde{\mathrm{Diff}}(X)/\mathrm{Diff}(X)$ measures the difference between the topological group $\mathrm{Diff}(X)$ of diffeomorphisms of the smooth manifold X and the simplicial group $\widetilde{\mathrm{Diff}}(X)$ of 'block diffeomorphisms', which is computable in terms of surgery theory, see [21]. Thus knowledge of the homotopy orbits for the involution acting on the spectrum $\mathrm{Wh}^{\mathrm{Diff}}(X)$, or equivalently on the spectrum $A(X)$, can be viewed as giving knowledge of the homotopy type of the space of diffeomorphisms $\mathrm{Diff}(X)$ in dimensions up to roughly n/3, where n is the dimension of X. Similar results apply for the spaces of PL homeomorphisms of PL manifolds and homeomorphisms of topological manifolds. See [63] for a more detailed survey.

In this paper we shall determine the homotopy type of the 2-primary completion of the spectrum $\mathrm{Wh}^{\mathrm{Diff}}(*)$. Since the Whitehead spectrum is a homotopy functor and preserves connectivity of maps, for any smooth n-manifold X which is roughly n/3-connected the map ϕ_X^{Diff} composed with the natural map

$$\Omega^\infty(EC_{2+} \wedge_{C_2} \Omega\mathrm{Wh}^{\mathrm{Diff}}(X)) \to \Omega^\infty(EC_{2+} \wedge_{C_2} \Omega\mathrm{Wh}^{\mathrm{Diff}}(*))$$

Is roughly n/3-connected. Thus, when our 2-primary calculation is extended to a calculation of the C_2-homotopy orbits of $\mathrm{Wh}^{\mathrm{Diff}}(*)$, we will have complete information about the 2-primary homotopy type of the space of diffeomorphisms $\mathrm{Diff}(X)$ of roughly n/3-connected manifolds up to dimension roughly n/3. We leave these calculations for a future paper.

Two-Primary Algebraic K-Theory of Pointed Spaces

We now turn to a description of the contents of the present paper.

We are able to access the homotopy type of $A(*)$ by means of a comparison of algebraic K-theory with the topological cyclic homology theory of Bökstedt et al. [8], relying on a theorem of Dundas [14]. In Section 1 we review these notions, and are led in Theorem 1.11 to the homotopy cartesian square

$$
\begin{array}{ccc}
A(*) & \xrightarrow{L} & K(\mathbb{Z}) \\
\downarrow{\scriptstyle trc_*} & & \downarrow{\scriptstyle trc_{\mathbb{Z}}} \\
TC(*) & \xrightarrow{L} & TC(\mathbb{Z}).
\end{array}
\tag{0.1}
$$

Here TC denotes the topological cyclic homology functor, and the natural transformation trc is the cyclotomic trace map of [8]. After 2-adic completion we can identify the homotopy type of $A(*)$ because the 2-primary homotopy types of the three other spectra in this diagram are known, together with sufficient information about the maps in the diagram. More specifically, the homotopy type of $TC(*)$ was determined in [8], for odd primes p the p-adic completion of $TC(\mathbb{Z})$ was computed in [9] and [10], and the 2-adic completion was determined in [45]. The 2-adic completion of $K(\mathbb{Z})$ was found in [46], by arguments based on Voevodsky's proof of the Milnor conjecture [51] and the Bloch–Lichtenbaum spectral sequence [4]. The 2-adic map $trc_{\mathbb{Z}} : K(\mathbb{Z}) \to TC(\mathbb{Z})$ was also studied in [45], in sufficient detail that we can describe $A(*)$ as an extension of $TC(*)$ by the common homotopy fiber of the maps labelled trc_* and $trc_{\mathbb{Z}}$ in the diagram above.

At odd primes p, the missing information needed to determine the p-primary homotopy type of $A(*)$ is the identification of the p-adic completion of $K(\mathbb{Z})$, i.e., a proof of the p-primary Lichtenbaum–Quillen conjecture for the integers, and the determination of how $A(*)$ is an extension of $TC(*)$ by the homotopy fiber of $trc_{\mathbb{Z}}$, after p-adic completion. Since $A(*)$ has finite type [17], and is rationally equivalent to $K(\mathbb{Z})$, this would suffice to determine the integral homotopy type of $A(*)$.

Also in Section 1 we make precise a part of the calculation of TC(∗) from [8], relating its p-adic completion to the Thom spectrum $\mathbb{C}P^{\infty}_{-1} = \text{Th}\left(-\gamma^{1}\right)$ of minus the canonical complex line bundle over $\mathbb{C}P^{\infty}$. See Theorem 1.16 and Corollary 1.21, which when combined yield a homotopy equivalence $TC(∗) \simeq \Sigma^{\infty} S^{0} \vee \Sigma \mathbb{C}P^{\infty}_{-1}$ after p-adic completion.

In Section 2 we analyze the 2-primary homotopy type of $\mathbb{C}P^{\infty}_{-1}$ by classical methods. We obtain its homotopy groups in dimensions $∗ \leq 20$ in Theorem 2.11, by use of the Atiyah–Hirzebruch spectral sequence for stable homotopy associated to the skeleton filtration of $\mathbb{C}P^{\infty}_{-1}$ by

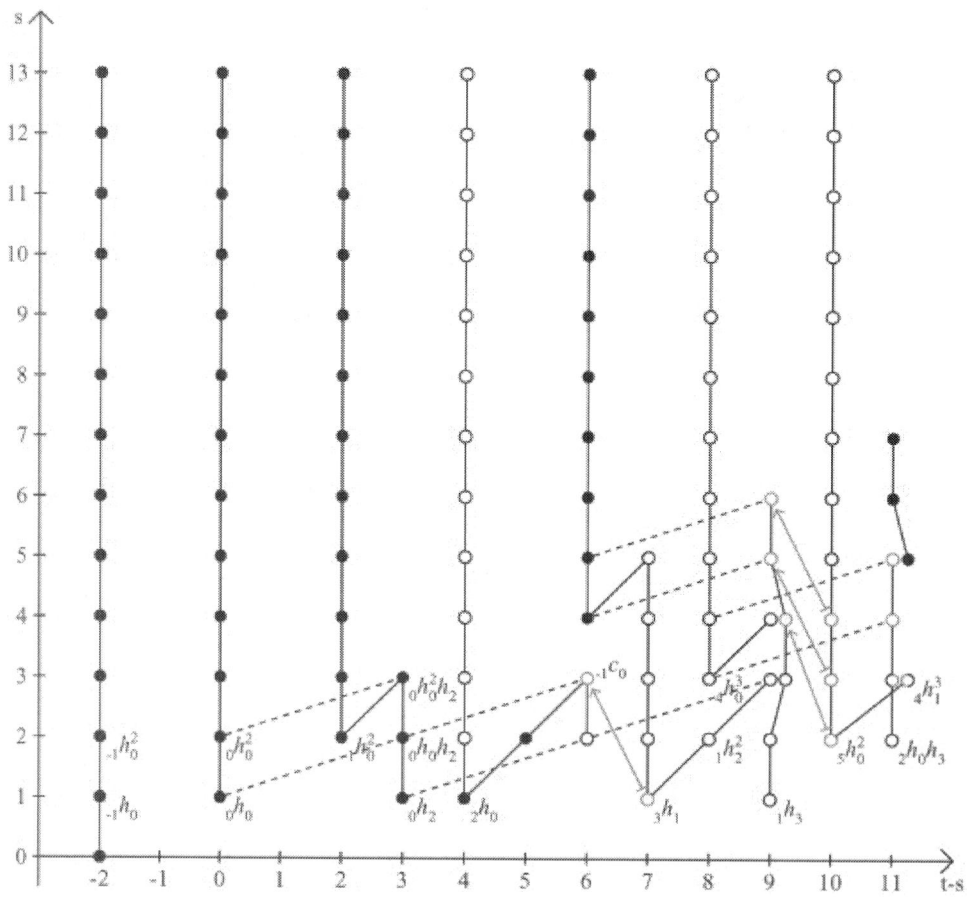

Table 3a: The Adams E_{2}-term for $\mathbb{C}P^{\infty}_{-1}$

the subspectra $\mathbb{C}P^s_{-1}$ for $s \geq -1$. The E^1-term in this spectral sequence is given in terms of the stable homotopy groups of spheres, π^S_*, and the differentials depend on the attaching maps for the cells in $\mathbb{C}P^\infty_{-1}$. This involves primary and secondary operations in homotopy, somewhat along the lines of Toda's book [50], and we build on previous work for $\mathbb{C}P^\infty$ by Mosher [34] and Mukai [35], [36] and [37].

It is much easier to describe $\mathbb{C}P^\infty_{-1}$ cohomologically, and in Proposition 2.13 we find that the mod 2 spectrum cohomology of $\mathbb{C}P^\infty_{-1}$ is cyclic as

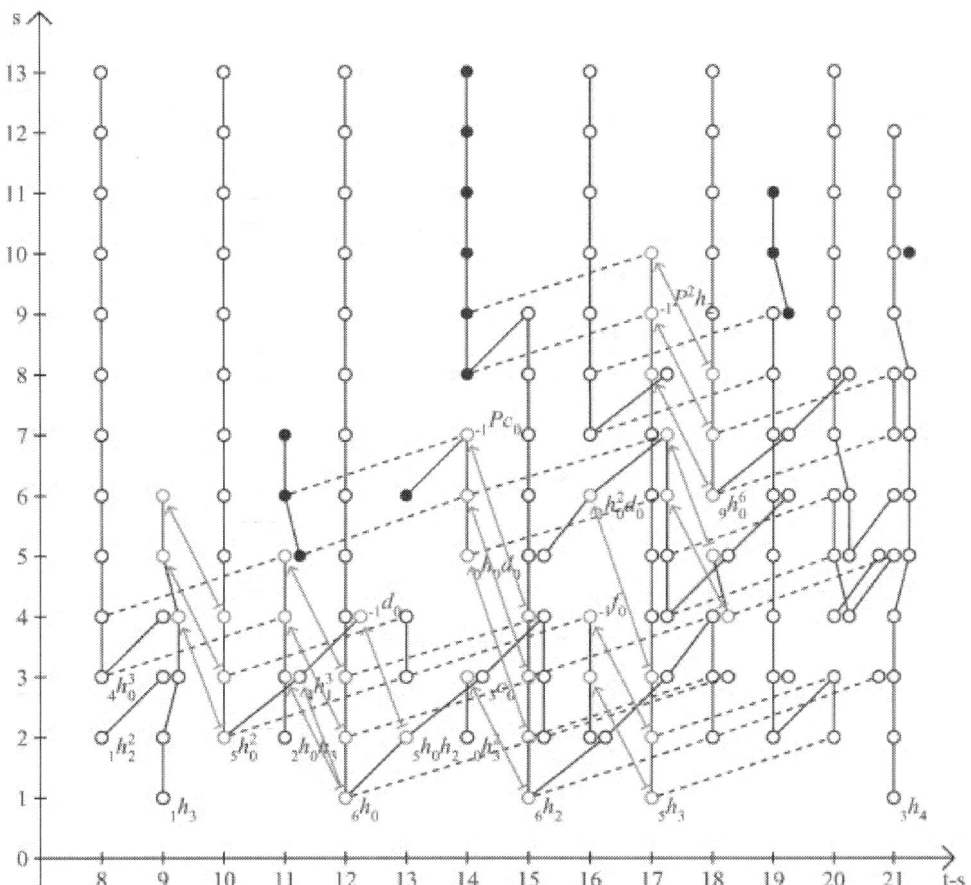

Table 3b: The Adams E_2-term for $\mathbb{C}P^\infty_{-1}$.

an A-module, where A is the mod 2 Steenrod algebra, and we describe the annihilator ideal C of the generator in Definition 2.12. The squaring operations Sq^i with i odd together with the admissible monomials Sq^I of length ≥ 2 form a basis for C as an $\mathbb{F}_2$-vector space. Thus $H^*_{spec}(\mathbb{C}P^\infty_{-1};$ $\mathbb{F}_2) \cong \Sigma^{-2}A/C$ as left graded A-modules. This allows us to describe the E_2-term of the Adams spectral sequence for the 2-adically completed homotopy of $\mathbb{C}P^\infty_{-1}$ in a range in Table 3a and Table 3b. Combined with the results from the Atiyah–Hirzebruch spectral sequence, we are also able to determine the differentials that land in homotopical degree t−s ≤ 20 in this spectral sequence. The details of this computation will be applied in Section 5, where Adams filtration and sparseness in the Adams spectral sequence will make it easier for us to study the homotopy type of A(∗) (and $Wh^{Diff}(*)$) in terms of its spectrum cohomology and the differentials in its Adams spectral sequence, rather than by means of the long exact sequence in homotopy arising from Dundas' homotopy cartesian square (0.1).

In Section 3 we familiarize ourselves with the spectrum hofib(trc) defined as the homotopy fiber of the (implicitly 2-completed) map

$$trc_{\mathbb{Z}} : K(\mathbb{Z}) \to TC(\mathbb{Z}).$$

By Dundas' theorem this is also the homotopy fiber of the map $trc_* : A(*) \to TC(*)$. The principal result is Theorem 3.13, which expresses this common homotopy fiber as the homotopy fiber of the spectrum map $\delta:\Sigma^{-2}ku \to \Sigma^4ko$ given as a suitably connected cover of the explicit composite map

$$\Sigma^4r \circ \beta^{-2} \circ (\psi^3 - 1) \circ \beta^{-1}:\Sigma^{-2}KU \to \Sigma^4KO.$$

From this description it is easy to extract other homotopical information about hofib(trc), such as its homotopy groups (Corollary 3.16), its spectrum cohomology (Theorem 4.4), or its Adams spectral sequence (Table 4a and Table 4b).

The calculations in Section 3 are based on the spectrum level description of $K\left(\mathbb{Z}\left[\frac{1}{2}\right]\right)$ given in Theorem 3.4, and of $K(\mathbb{Q}_2)$ given in Theorem 3.6, which were obtained in [46] and [45, 8.1], respectively. The calculation of $K\left(\mathbb{Z}\left[\frac{1}{2}\right]\right)$ relied on the proven Lichtenbaum–Quillen conjecture in this case [46], using essential inputs from algebraic geometry [4] and [51], while the identification of $K(\mathbb{Q}_2)$ in [45] amounted to the calculation of $TC(\mathbb{Z})$ completed at 2, which used topological cyclic homology and calculational spectral sequence techniques from stable homotopy theory. The results in Section 3 also rely on knowing how the natural map $j' : K\left(\mathbb{Z}\left[\frac{1}{2}\right]\right) \to K(\mathbb{Q}_2)$ acts on the level of homotopy groups, which was determined in [45, 7.7 and 9.1]. Those results depended on knowing the structure of the K-theory spectra involved, not just their homotopy groups, and were feasible because the prime 2 is so small, or perhaps because it is regular.

These inputs allow us to obtain a spectrum level description of the homotopy fiber of j' in Proposition 3.10 and Proposition 3.11, with a more convenient reformulation given in Proposition 3.12. The arguments rely on knowing the endomorphism algebras of the 2-completed connective topological K-theory spectra ko and ku, as well as all the maps between them, which stems from [29]. Using Quillen's localization sequence in algebraic K-theory, and Hesselholt and Madsen's link between $K(\mathbb{Z}_2)$ and $TC(\mathbb{Z})$ from [22, Theorem D], we rework the description of hofib(j') into a spectrum level description of hofib(trc) in Theorem 3.13, as desired.

In Section 4 we use the cofiber sequence (3.14)

$$\mathbb{C}P^{\infty}_{-1} \xrightarrow{i} \text{hofib}(\text{trc}) \xrightarrow{j} \text{Wh}^{\text{Diff}}(*)$$

And the splitting $A(*) \simeq \Sigma^{\infty} S^0 \vee \mathrm{Wh}^{\mathrm{Diff}}(*)$, to reduce the identification of $A(*)$ to that of $\mathbb{CP}^{\infty}_{-1}$, which was studied in Section 2, to that of hofib(trc), which was settled in Section 3, and the map ibetween the two. At the prime 2 we are in the fortunate situation that the mod2 spectrum cohomology of $\mathbb{CP}^{\infty}_{-1}$ is cyclic as an A-module on a generator in degree -2, so because $\mathrm{Wh}^{\mathrm{Diff}}(*)$ is 2-connected it follows that i induces a surjection on cohomology in all degrees. Thus we can omit any discussion of the linearization map $L: TC(*) \to TC(\mathbb{Z})$ in Dundas' homotopy cartesian square, and still obtain a complete cohomological description of $\mathrm{Wh}^{\mathrm{Diff}}(*)$.

This is achieved in the main Theorem 4.5. We have an isomorphism of left graded A-modules

$$H^*_{\mathrm{spec}}(A(*); \mathbb{F}_2) \cong H^*_{\mathrm{spec}}(\Sigma^{\infty} S^0; \mathbb{F}_2) \oplus H^*_{\mathrm{spec}}(\mathrm{Wh}^{\mathrm{Diff}}(*); \mathbb{F}_2),$$

Where

$$H^*_{\mathrm{spec}}(\Sigma^{\infty} S^0; \mathbb{F}_2) = \mathbb{F}_2$$

Is the trivial A-module in dimension zero, and there is a unique nontrivial extension of left graded A-modules

$$\Sigma^{-2} C/A(Sq^1, Sq^3) \to H^*_{\mathrm{spec}}(\mathrm{Wh}^{\mathrm{Diff}}(*); \mathbb{F}_2) \to \Sigma^3 A/A(Sq^1, Sq^2)$$

Characterizing $H^*_{\mathrm{spec}}(\mathrm{Wh}^{\mathrm{Diff}}(*); \mathbb{F}_2)$. Here $C \subset A$ is the annihilator ideal of the generator for $H^*_{\mathrm{spec}}(\mathbb{CP}^{\infty}_{-1}; \mathbb{F}_2)$, introduced in Definition 2.12. The assertion of the theorem is that algebraically there are precisely two such extensions of left graded A-modules, and $H^*_{\mathrm{spec}}(\mathrm{Wh}^{\mathrm{Diff}}(*); \mathbb{F}_2)$. is the one that does not split.

In Section 5 we turn to a homotopical analysis of the smooth White-head spectrum $\text{Wh}^{\text{Diff}}(*)$, and thus also of $A(*)$. Our approach is to study the Adams spectral sequence (5.4)

$$E_2^{s,t} = \text{Ext}_A^{s,t}(H_{\text{spec}}^*(\text{Wh}^{\text{Diff}}(*); \mathbb{F}_2), \mathbb{F}_2) \Rightarrow \pi_{t-s}(\text{Wh}^{\text{Diff}}(*))_2^{\wedge}.$$

Here we can, in principle, compute the E_2-term in a large range of bi-degrees, but there will be many families of differentials and a complete determination of the homotopy groups of $\text{Wh}^{\text{Diff}}(*)$ is out of reach.

The cofiber sequence (3.14) displayed above has the special proper-ty that its connecting map induces the zero map in mod 2 spectrum cohomology, so its associated long exact sequence in cohomology breaks up into short exact sequences, which in turn induce long ex-act sequences of ExtA-groups. Thus the E_2-terms of the Adams spec-tral sequences for $\mathbb{C}P_{-1}^{\infty}$, hofib (trc) and $\text{Wh}^{\text{Diff}}(*)$ are linked in a long exact sequence (5.5). The spectral sequence for hofib(trc) was com-pletely described in Section 3, and inSection 5 we use the long exact sequence of E_2-terms to translate the information from Section 2 about differentials in the Adams spectral sequence for $\mathbb{C}P_{-1}^{\infty}$ to information about differentials in the Adams spectral sequence (5.5) for $\text{Wh}^{\text{Diff}}(*)$. This is a convenient approach, because the Adams spectral sequence of hofib(trc) is concentrated above the line t−s=2s+3, while the dif-ferentials in the spectral sequence for $\mathbb{C}P_{-1}^{\infty}$ mostly originate below this line. The only subtle point concerns whether certain h_1-divisible classes in bidegrees (s,t)=(4k,12k+3) of (5.4) are hit by differentials, but a comparison with [45, 9.1] reveals that they indeed do survive to the E_∞-term. Thus the complexity of determining the homotopy groups of $\text{Wh}^{\text{Diff}}(*)$ is in practice equivalent to that of determining the homotopy groups of $\mathbb{C}P_{-1}^{\infty}$, which is a well explored but not exhaustively analyzed problem.

The Adams E_2-term for $\text{Wh}^{\text{Diff}}(*)$ is displayed in part in Table 6a and Ta-ble 6b, and the non-zero differentials landing in homotopical dimen-sion t−s≤21 are listed in Proposition 5.7. This leads to a calculational

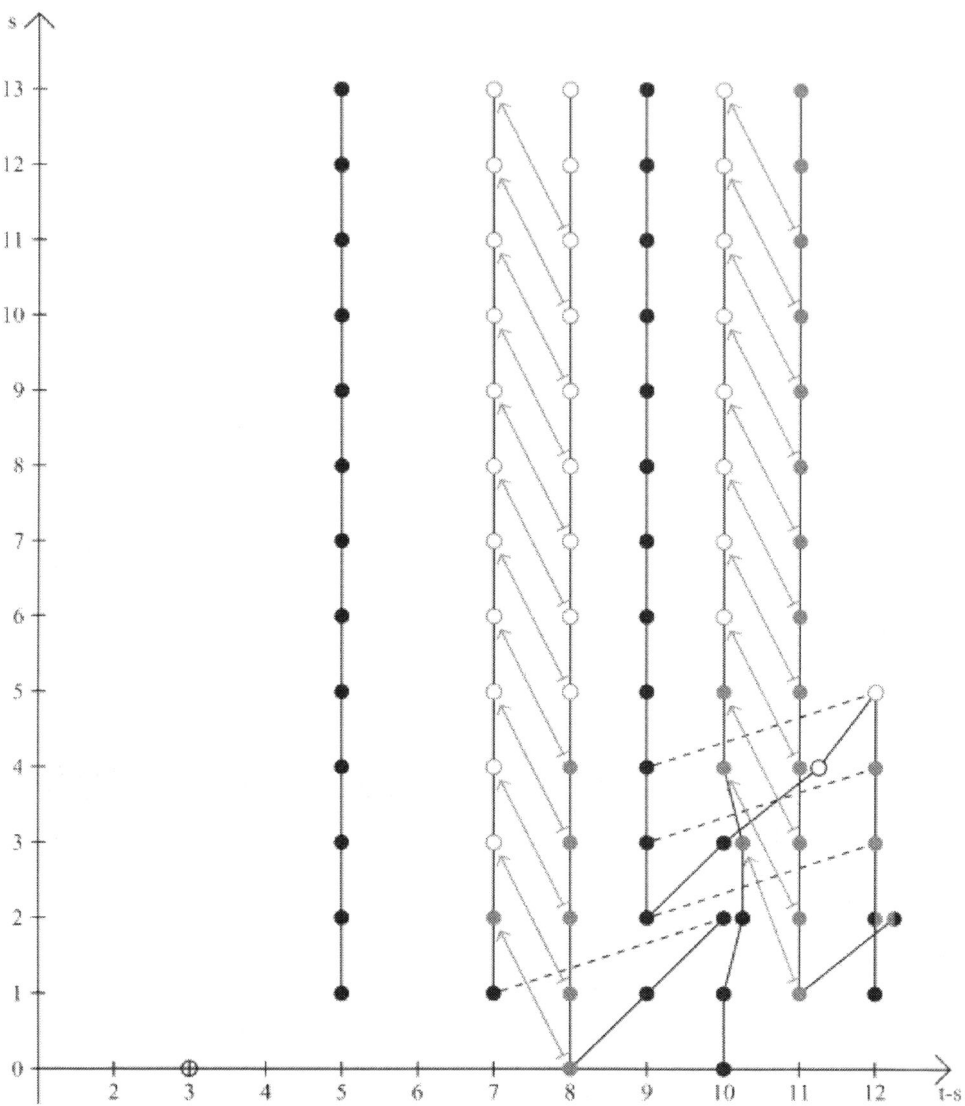

Table 6a: The Adams E_2-term for $\mathrm{Wh}^{\mathrm{Diff}}(*)$

conclusion in Theorem 5.8, where the 2-completed homotopy groups of $\mathrm{Wh}^{\mathrm{Diff}}(*)$ are listed in dimensions $* \leq 18$, and up to group extensions in dimensions $19 \leq * \leq 21$. Previously, only the homotopy groups in dimensions ≤ 3 were known, see [11]. We do not give names to the classes identified in $\pi_*\left(\mathrm{Wh}^{\mathrm{Diff}}(*)\right)$, but in Theorem 7.5 we show that

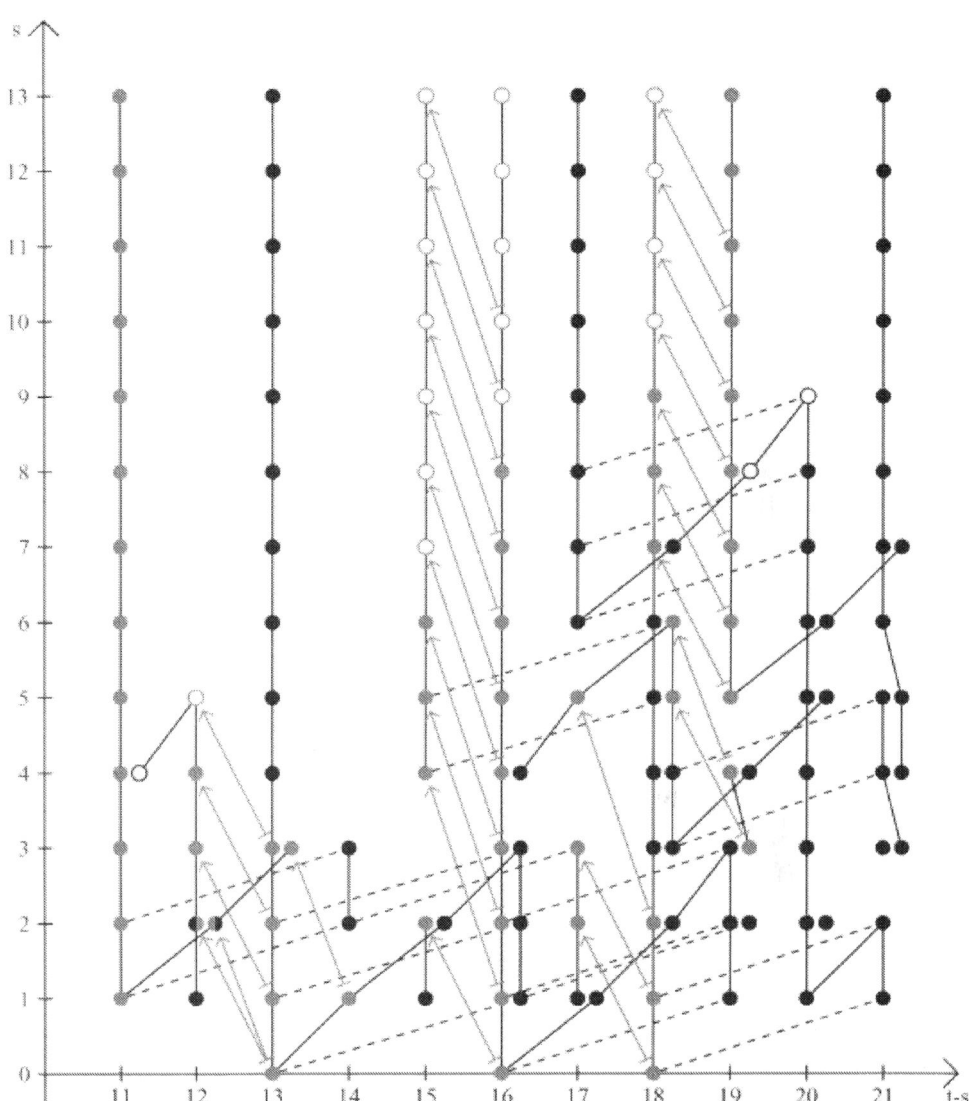

Table 6b: The Adams E_2-term for $\mathrm{Wh}^{\mathrm{Diff}}(*)$

the (space level) Hatcher–Waldhausen map $\mathrm{hw}: G/O \to \mathrm{Wh}^{\mathrm{Diff}}(*)$ constructed in [55, Section 3] induces an isomorphism on 2-primary homotopy groups in dimensions $* \leq 8$, and an injection on 2-primary homotopy groups in dimensions $* \leq 13$. Thus, the better known homotopy groups of $G/O \simeq BSO \times Cok\, J$ account for much of the low-dimensional homotopy of $Wh^{Diff}(*)$.

In Section 6 we use the known spectrum level description of $K(\mathbb{Z})$ completed at 2 to compute its mod 2 spectrum cohomology in Theorem 6.4, and to show in Corollary 6.7 that the linearization map

$L : A(*) \to K(\mathbb{Z})$ induces the zero map in mod 2 spectrum cohomology in positive dimensions. Thus the linearization map does not itself provide a good cohomological approximation to $A(*)$. In Remark 6.8 we explain why the Hatcher–Waldhausen map hw does not admit a four-fold delooping, using that multiplication by the Hopf map $\sigma \in \pi_7^S$

is non-zero on $\pi_4\big(\Omega Wh^{Diff}(*)\big)$, but is zero on $\pi_4(G/O)$. We also explain how this relates to the results of [41], where an infinite loop map

from G/O to a different infinite loop space structure on $\Omega Wh^{Diff}(*)$ was obtained.

Following Miller and Priddy [32], we describe in (6.3) a spectrum $g/o_\oplus$

as the homotopy fiber of the 2-completed unit map $\Sigma^\infty S^0 \to K(\mathbb{Z})$. Its underlying space $G/O_\oplus$ has the same 2-adic homotopy type as the usual

G/O. Although there is no spectrum map $\Sigma g/o_\oplus \to Wh^{Diff}(*)$ inducing a π_3-isomorphism, we construct in Section 7 a 2-complete spectrum

map $M : Wh^{Diff}(*) \to \Sigma g/o_\oplus$ which induces an isomorphism on mod 2 spectrum cohomology in all dimensions $* \le 9$. This is a best possible approximation, since the cohomology groups differ in dimension 10.

The comparison of $Wh^{Diff}(*)$ with $\Sigma g/o_\oplus$ finally allows us to evaluate the Hatcher–Waldhausen map on 2-completed homotopy groups in dimensions $* \le 13$, leading to the previously cited Theorem 7.5.

ALGEBRAIC K-THEORY AND TOPOLOGICAL CYCLIC HOMOLOGY

We commence by discussing the cyclotomic trace map from algebraic K-theory to topological cyclic homology, and a special case of Dundas' theorem comparing relative algebraic K-theory to relative topological cyclic homology.

Γ-spaces and $\mathbb{S}$-algebras

Let $\mathscr{S}_*$ be the category of pointed simplicial sets, and let Γ^{op} be the category of finite pointed sets $k_+ = \{0,1,\ldots,k\}$ based at 0, and base-point preserving functions. This is the opposite of Segal's category Γ from [47]. Let $\Gamma\mathscr{S}_*$ be the category of Γ-spaces, i.e., functors $F: \Gamma^{op} \to \mathscr{S}_*$ with $F(0_+) = *$. Each Γ-space F naturally extends to a functor F: $\mathscr{S}_* \to \mathscr{S}_*$, which when evaluated on spheres determines a connective (pre-) spectrum $\{n \mapsto F(S^n)\}$. We write $\pi_*(F)$ for the homotopy groups of this spectrum. The natural inclusion $\Gamma^{op} \to \mathscr{S}_*$ is a Γ-space denoted $\mathbb{S}$, whose associated spectrum is the sphere spectrum. The groups $\pi_*(\mathbb{S})$ are the stable homotopy groups of spheres.

There is a smash product $\wedge$ of Γ-spaces defined by Lydakis [27], making $(\Gamma\mathscr{S}_*, \wedge, \mathbb{S})$ a symmetric monoidal category. A monoid A in this symmetric monoidal category will be called an $\mathbb{S}$-algebra. Its associated spectrum is an associative ring spectrum, conveniently thought of as an algebra over the sphere spectrum.

Examples of $\mathbb{S}$-algebras

When G is a simplicial group the functor $\Sigma^\infty(G_+)$ given by $\Sigma^\infty(G_+)(k_+) = G_+ \wedge k_+$ is a Γ-space. The group multiplication and unit define the structure maps

$$\mu: \Sigma^\infty(G_+) \wedge \Sigma^\infty(G_+) \to \Sigma^\infty(G_+)$$

and $\eta : \mathbb{S} \to \Sigma^\infty(G_+)$ making $\Sigma^\infty(G_+)$ an $\mathbb{S}$-algebra. Its associated ring spectrum is the unreduced suspension spectrum on G, with product map induced by the multiplication on G.

When R is a (discrete) ring the functor HR given by $HR(k_+)=R\{1,...,k\}$ (the free R-module on the non-basepoint elements in k_+) is a Γ-space. The ring multiplication and unit define the structure maps

$$\mu : HR \wedge HR \to HR$$

and $\eta : \mathbb{S} \to HR$ making HR an $\mathbb{S}$-algebra. Its associated ring spectrum is the Eilenberg–Mac Lane spectrum representing ordinary cohomology with coefficients in R.

Let G be a simplicial group, with group of path components $\pi_0(G)$, and let $R = \mathbb{Z}[\pi_0(G)]$ be its integral group ring. The linearization map is the map of $\mathbb{S}$-algebras $L : \Sigma^\infty(G_+) \to HR$ taking $g \wedge i \in G_+ \wedge k_+$ to $[g] \cdot i \in R\{1,...,k\}$, where $g \in G, i \in \{1,...,k\}$ and [g] denotes the path component of g viewed as an element of $\pi_0(G) \subset R$.

Algebraic k-theory, Topological Hochschild Homology And Topological Cyclic Homology

Let A be an $\mathbb{S}$-algebra. The extended functor $A : \mathscr{S}_* \to \mathscr{S}_*$ comes equipped with a product and unit map making it an FSP (functor with smash product) in the sense of [6]. In [8] Bökstedt, Hsiang and Madsen functorially define the algebraic K-theory spectrum K(A), topological Hochschild homology spectrumTHH(A) and topological cyclic homology spectrum TC(A,p) of an FSP A. Here p is any prime. An integral functor A↦TC(A) has been defined by Goodwillie (unpublished), together with a natural p-adic equivalenceTC(A)→TC(A,p) for each prime p.

When G is a simplicial group with classifying space X=BG, we write $A(X)=K(\Sigma^\infty(G_+))$, $THH(X)=THH(\Sigma^\infty(G_+))$ and $TC(X,p)=TC(\Sigma^\infty(G_+),p)$. Here A(X) is naturally homotopy equivalent to Waldhausen's algebraic K-theory spectrum A(X) of the space X [53], i.e., the algebraic K-theory of the category of finite retractive spaces over X.

When R is a ring we write K(R)=K(HR), THH(R)=THH(HR) and TC(R,p)=TC(HR,p). Here K(R) is naturally homotopy equivalent to Quillen's algebraic K-theory spectrum K(R) of the ring R [39], i.e., the algebraic K-theory of the category of finitely generated projective R-modules.

We recall from [8, 3.7] that there are C-equivariant homotopy equivalences

$$THH(X) \simeq_C \Sigma_C^\infty(\Lambda X_+)$$

for each finite subgroup $C \subset S^1$. Here Σ_C^∞ denotes the C-equivariant suspension spectrum, and $C \subset S^1$ acts on the free loop space ΛX by rotating the loops.

Trace Maps

A trace map $tr_X:A(X) \to THH(X)$ was defined by Waldhausen [54], and Bökstedt defined a trace map $tr_A:K(A) \to THH(A)$ in [6], as a natural transformation of functors from FSPs to spectra. The cyclotomic trace map trcA of [8] gives a factorization

$$K(A) \xrightarrow{trc_A} TC(A, p) \xrightarrow{\beta_A} THH(A)$$

of tr_A, although the map to TC(A,p) was initially only defined up to homotopy. The map β_A is a projection map from the homotopy limit defining TC(A,p). When $A=\Sigma^\infty(G_+)$ with X=BG or A=HR we substitute X or R, respectively, for A in the notations trc_A, β_A and tr_A. Thus $trc_X:A(X) \to TC(X,p)$, etc.

In the case $A=\Sigma^\infty(G_+)$ with X=BG the six authors of [7] gave a model for the cyclotomic trace map trcX as a natural transformation in X. When A=HR, Dundas and McCarthy [16] gave models for K(R) and TC(R) such that trc_R is a natural transformation. Finally, Dundas [15] has provided a construction of functors K, THHand TC from $\mathbb{S}$-algebras to spectra, and natural transformations trc:K→TC, β:TC→THH and tr:K→THH with tr=β∘trc, which agree up to natural homotopy equivalence with the preceding definitions.

Dundas' Theorem

The following theorem of Dundas [14] generalizes to maps of $\mathbb{S}$-algebras a theorem of McCarthy [31] valid for maps of simplicial rings. Both results are analogous to an older theorem about rational algebraic K-theory due to Goodwillie [20].

Theorem 1.7 Dundas

Letφ:A→Bbe a map of $\mathbb{S}$-algebras, such that the ring homomorphism$\pi_0(\varphi):\pi_0(A)\to\pi_0(B)$ is a surjection with nilpotent kernel. Then the commutative square of spectra

$$
\begin{array}{ccc}
K(A) & \xrightarrow{\phi} & K(B) \\
\downarrow{\scriptstyle trc_A} & & \downarrow{\scriptstyle trc_B} \\
TC(A) & \xrightarrow{\phi} & TC(B)
\end{array}
$$

is homotopy cartesian.

Corollary 1.8 Dundas

Let G be a simplicial group, and write X=BG and $R = \mathbb{Z}[\pi_1(X)] = \mathbb{Z}[\pi_0(G)]$. The linearization map $L:\Sigma^\infty(G_+)\to HR$ induces a homotopy cartesian square

$$
\begin{array}{ccc}
A(X) & \xrightarrow{L} & K(R) \\
\downarrow{\scriptstyle trc_X} & & \downarrow{\scriptstyle trc_R} \\
TC(X) & \xrightarrow{L} & TC(R).
\end{array}
$$

In particular, the vertical homotopy fiber hofib(trc$_X$) only depends on the fundamental group$\pi_1(X)$, for a pointed connected space X.

For the last claim we used that any pointed connected space X is homotopy equivalent to BG for a simplicial group G, e.g. the Kan loop group of X. See [59].

Whitehead Spectra

There are natural cofiber sequences of spectra

$$\Sigma^\infty(X_+) \xrightarrow{i} A(X) \to \mathrm{Wh}^{\mathrm{Diff}}(X)$$

and

$$A(*) \wedge X_+ \xrightarrow{\alpha} A(X) \to \mathrm{Wh}^{\mathrm{PL}}(X),$$

where Wh^{Diff}(X) is the smooth Whitehead spectrum of X, and Wh^{PL}(X) is the piecewise linear Whitehead spectrum of X. The sequences are constructed geometrically in [55], where Wh^{Diff}(X) is interpreted in terms of stabilized smooth concordance spaces and stabilized spaces of smooth h-cobordisms, and similarly in the piecewise linear case. The identification of the upper left hand homology theory in X with$\Sigma^\infty(X_+)$ uses the 'vanishing of the mystery homology theory' established in [58].

The composite

$$\Sigma^\infty(X_+) \xrightarrow{i} A(X) \xrightarrow{\mathrm{tr}_X} THH(X) \simeq \Sigma^\infty(\Lambda X_+) \xrightarrow{ev} \Sigma^\infty(X_+)$$

is homotopic to the identity. Here ev:$\Lambda X \to X$ is the map evaluating a free loop $S^1 \to X$ at the identity $1 \in S^1$. Hence ev$\circ$tr$_X$ provides a natural splitting for the first cofiber sequence above, as in

$$A(X) \simeq \Sigma^\infty(X_+) \vee \mathrm{Wh}^{\mathrm{Diff}}(X).$$

We can therefore identify Wh^{Diff}(X) with the homotopy fiber of the splitting map ev$\circ$tr$_X$.

The Smooth Whitehead Spectrum of A Point

Suppose $G=1$, so that $X=*$. Then $ev: \Lambda X \to X$ is the identity map, $THH(*) \simeq \Sigma^\infty S^0$, and the splitting above identifies $Wh^{Diff}(*)$ with the homotopy fiber of tr_*. We obtain a map of horizontal (co-)fiber sequences of spectra:

$$
\begin{array}{ccccc}
Wh^{Diff}(*) & \longrightarrow & A(*) & \xrightarrow{tr_*} & THH(*) \\
\downarrow{\scriptstyle \widetilde{trc}} & & \downarrow{\scriptstyle trc_*} & & \| \\
\widetilde{TC}(*) & \longrightarrow & TC(*) & \xrightarrow{\beta_*} & THH(*).
\end{array}
$$

Here $\widetilde{TC}(*)$ is defined as the homotopy fiber of β_*, and $\widetilde{trc}$ is the induced map of homotopy fibers over trc_* and the identity map on THH $(*)$. The unit map $\Sigma^\infty S^0 \to A(*) \to TC(*)$ and β_* yield a splitting

$$TC(*) \simeq \Sigma^\infty S^0 \vee \widetilde{TC}(*).$$

Theorem 1.11

The two squares

$$
\begin{array}{ccccc}
Wh^{Diff}(*) & \longrightarrow & A(*) & \xrightarrow{L} & K(\mathbb{Z}) \\
\downarrow{\scriptstyle \widetilde{trc}} & & \downarrow{\scriptstyle trc_*} & & \downarrow{\scriptstyle trc_{\mathbb{Z}}} \\
\widetilde{TC}(*) & \longrightarrow & TC(*) & \xrightarrow{L} & TC(\mathbb{Z})
\end{array}
$$

are homotopy cartesian, and induce homotopy equivalences of vertical homotopy fibers

$$\text{hofib}(\widetilde{trc}) \xrightarrow{\simeq} \text{hofib}(trc_*) \xrightarrow{\simeq} \text{hofib}(trc_{\mathbb{Z}}).$$

We denote either of these by hofib(trc).

The Topological Cyclic Homology of A Point

The topological cyclic homology $TC(X,p)$ of a pointed connected space X was computed by Bökstedt et al.[8]. We recall their result, making precise a point that was omitted in the published argument. See [28, Section 4.4] for more details about the following review.

Fix a prime p. From (1.4) there is an equivalence $THH(X)^{C_{p^n}} \simeq \sum_{C_{p^n}}^\infty (\wedge X_+)^{C_{p^n}}$ for each $n<0$. The Segal–tom Dieck splitting

$$\sum_{C_{p^n}}^\infty (\Lambda X_+)^{C_{p^n}} \simeq \prod_{k=0}^{n} \Sigma^\infty (EC_{p^k} \times_{C_{p^k}} \Lambda X^{C_{p^{n-k}}})_+$$

and the power map homeomorphisms $\Delta_p^{n-k} : \Lambda X \cong \Lambda X^{C_{p^{n-k}}}$ combine to give an equivalence

$$THH(X)^{C_{p^n}} \simeq \prod_{k=0}^{n} \Sigma^\infty (EC_{p^k} \times_{C_{p^k}} \Lambda X)_+.$$

$$(1.13)$$

The pth power map $\Delta p : \Sigma^\infty \Lambda X_+ \to \Sigma^\infty \Lambda X_+$ is induced by taking a free loop $S^1 \to X$ to its precomposition by the standard degree p map $S^1 \to S^1$. Let

$$t_p : \Sigma^\infty (EC_{p^n} \times_{C_{p^n}} \Lambda X)_+ \to \Sigma^\infty (EC_{p^{n-1}} \times_{C_{p^{n-1}}} \Lambda X)_+$$

be the Becker–Gottlieb transfer for the principal C_p-bundle $EC_{p^{n-1}} \times_{C_{p^{n-1}}} \Lambda X \to EC_{p^n} \times_{C_{p^n}} \Lambda X$. There are restriction and Frobenius maps R,F: $THH(X)^{C_{p^n}} \to THH(X)^{C_{p^{n-1}}}$. Up to homotopy these are given by the formulas:

$$R(x_0, x_1, \ldots, x_n) = (x_0, x_1, \ldots, x_{n-1}),$$

$$F(x_0, x_1, \ldots, x_n) = (\Delta_p(x_0) + t_p(x_1), t_p(x_2), \ldots, t_p(x_n)).$$

Here x_k refers to the factor in $\Sigma^\infty(EC_{p^k} \times_{C_{p^k}} \Lambda X)_+$ in the equivalence (1.13), and the formulas must be interpreted as giving maps defined in terms of this splitting.

Writing

$$TR(X, p) = \operatorname*{holim}_{n,R} THH(X)^{C_{p^n}} \simeq \prod_{n=0}^{\infty} \Sigma^\infty(EC_{p^n} \times_{C_{p^n}} \Lambda X)_+,$$

$$(1.14)$$

we have $R(x_0,x_1,x_2,\ldots)=(x_0,x_1,x_2,\ldots)$ and $F(x_0,x_1,x_2,\ldots)=(\Delta p(x_0)+tp(x_1),tp(x_2),tp(x_3),\ldots)$ up to homotopy. The topological cyclic homology spectrum $TC(X,p)$ is defined as the homotopy equalizer

$$TC(X, p) \xrightarrow{\pi} TR(X, p) \underset{F}{\overset{R}{\rightrightarrows}} TR(X, p)$$

And is homotopy equivalent to the homotopy fiber of $1-F$: $TR(X,p) \to TR(X,p)$. Let $T, D: TR(X,p) \to TR(X,p)$ be given up to homotopy by the formulas:

$$T(x_0,x_1,x_2,\ldots) = (t_p(x_1), t_p(x_2), t_p(x_3),\ldots),$$

$$D(x_0,x_1,x_2,\ldots) = (\Delta_p(x_0), 0, 0,\ldots).$$

The following observation lets us calculate $TC(X,p)$.

Lemma 1.15

The composite $(1-T) \circ (1-D): TR(X,p) \to TR(X,p)$ is homotopic to $(1-F)$.

Proof

In terms of the splitting (1.14) it is clear that $(1-D)(x_0,x_1,x_2,\ldots)=(x_0-\Delta p(x_0),x_1,x_2,\ldots)$ is mapped by $(1-T)$ to $(x_0-\Delta p(x_0)-tp(x_1),x_1-tp(x_2),x_2-tp(x_3),\ldots)$, which is homotopic to $(1-F)(x_0,x_1,x_2,\ldots)$.

Given such a choice of commuting homotopy for the right hand square below, there is an induced map of horizontal (co-)fiber sequences

$$
\begin{array}{ccccc}
TC(X, p) & \xrightarrow{\pi} & TR(X, p) & \xrightarrow{1-F} & TR(X, p) \\
\downarrow{\alpha_X} & & \downarrow{1-D} & & \| \\
C(X, p) & \longrightarrow & TR(X, p) & \xrightarrow{1-T} & TR(X, p).
\end{array}
$$

Here we have written C(X,p) for the homotopy limit $\operatorname{holim}_{n,tp} \Sigma^\infty (EC_{p^n} \times_{C_{p^n}} \Lambda X)_+$, which is homotopy equivalent to the homotopy fiber of $1-T$ in view of (1.14). When α_X is determined by the right hand commuting homotopy, the left hand square is homotopy cartesian (with respect to a suitable commuting homotopy). Let $pr:TR(X,p)\to THH(X)\simeq\Sigma^\infty\Lambda X_+$ denote projection to the zeroth term in the homotopy limit defining TR(X,p). Then there is clearly a homotopy cartesian square

$$
\begin{array}{ccc}
TR(X, p) & \xrightarrow{pr} & \Sigma^\infty \Lambda X_+ \\
\downarrow{1-D} & & \downarrow{1-\Delta_p} \\
TR(X, p) & \xrightarrow{pr} & \Sigma^\infty \Lambda X_+,
\end{array}
$$

Again with respect to a suitable commuting homotopy. We can combine these two homotopy cartesian squares horizontally. Then the upper composite $\beta_X=pr\circ\pi: TC(X,p)\to TR(X,p)\to THH(X)\simeq\Sigma^\infty\Lambda X_+$ agrees with the natural transformation β of 1.5. The lower composite is the projection $pr_0: C(X,p)\to\Sigma^\infty\Lambda X_+$ from the homotopy limit system over the Becker–Gottlieb transfer maps to its zeroth term.

Theorem 1.16 Bökstedt et al. [8, Section 5]

Let X be a pointed connected space and write $C(X,p)=\operatorname{holim}_{n,tp}\Sigma^\infty(EC_{p^n}\times_{C_{p^n}}\Lambda X)_+$. The diagram

$$TC(X, p) \xrightarrow{\alpha_X} C(X, p)$$

$$\downarrow \beta_X \qquad\qquad \downarrow pr_0$$

$$\Sigma^\infty \Lambda X_+ \xrightarrow{1-\Delta_p} \Sigma^\infty \Lambda X_+$$

homotopy commutes, and there exists a commuting homotopy making the diagram homotopy cartesian.

This is now clear. (The proofs in [8] and [28] only show that the horizontal homotopy fibers in this diagram are homotopy equivalent, not necessarily by the map induced by β_X and pr_0.) Specializing to X=* we have the following corollary, which is what we will use in the rest of the present paper.

Corollary 1.17

There is a (co-)fiber sequence of spectra

$$\widetilde{TC}(*, p) \to \underset{n, t_p}{\mathrm{holim}}\, \Sigma^\infty(BC_{p^n+}) \xrightarrow{pr_0} \Sigma^\infty S^0.$$

For each $n \geq 0$ there is a dimension-shifting S^1-transfer map

$$\mathrm{trf}^n_{S^1} : \Sigma^\infty(\Sigma(\mathbb{C}P^\infty_+)) \to \Sigma^\infty(BC_{p^n+})$$

Associated to the S^1-bundle $BC_{p^n} \to BS^1 \simeq \mathbb{C}P^\infty$. See [25], [26] and [35]. These induce a map

$$\Sigma^\infty(\Sigma(\mathbb{C}P^\infty_+)) \to \underset{n, t_p}{\mathrm{holim}}\, \Sigma^\infty(BC_{p^n+})$$

Which is a homotopy equivalence after p-adic completion. Hence, we can identify the map pr_0 above with the S^1-transfer map $\mathrm{trf}S^{10}$, briefly denoted trf_s^1, after p-adic completion. Combined with the p-adic equivalence $TC(*) \to TC(*, p)$ we obtain:

Corollary 1.18 Bökstedt et al

There is a homotopy equivalence

$$\widetilde{TC}(*) \simeq \mathrm{hofib}(\mathrm{trf}_{S^1} : \Sigma^\infty(\Sigma(\mathbb{C}P_+^\infty)) \to \Sigma^\infty S^0)$$

after p-adic completion, for each prime p.

A Thom spectrum

Let $\mathbb{C}P_k^\infty$ denote the truncated complex projective space with one cell in each even dimension greater than or equal to $2k$, interpreted as a spectrum when $k<0$. There is a homotopy equivalence

$$\mathbb{C}P_k^\infty \simeq Th(k\gamma^1),$$

Where the right hand side is the Thom spectrum of k times the canonical complex line bundle over $\mathbb{C}P^\infty$ see [3]. We shall be concerned with the case $k=-1$, i.e., with the spectrum $\mathbb{C}P_{-1}^\infty$, which can be thought of as the Thom spectrum of minus the canonical line bundle on $\mathbb{C}P^\infty$.

Theorem 1.20 Knapp

There is a homotopy equivalence

$$\Sigma\mathbb{C}P_{-1}^\infty \simeq \mathrm{hofib}(\mathrm{trf}_{S^1} : \Sigma^\infty(\Sigma(\mathbb{C}P_+^\infty)) \to \Sigma^\infty S^0).$$

See [25, 2.9] for a proof. Bringing these results together we have shown:

Corollary 1.21

There is a homotopy equivalence

$$(\Sigma\mathbb{C}P_{-1}^\infty)_p^\wedge \simeq \widetilde{TC}(*)_p^\wedge$$

of p-adically completed spectra.

Table 1: E^4 in total degrees $s+t \leq 20$

0											
0	$4(2\bar{\kappa}_0)$										
$2(\bar{\kappa}_{-1})$	$2(\bar{\sigma}_0)$	$2(2\nu_1^*)$									
	$\oplus$ $8(\bar{\zeta}_0)$										
$4(\bar{\zeta}_{-1})$	$8(\nu_0^*)$	$2(\bar{\mu}_1)$	0	$32(\rho_3)$							
$4(\nu_{-1}^*)$	$2(\nu\kappa_0)$ $\oplus$ $2(\eta^2\rho_0)$	$2(\eta_1^*)$	$16(2\rho_2)$	$2(\sigma_3^2)$ $\oplus$ $2(\kappa_3)$	0						
$2(\bar{\mu}_{-1})$	0	$32(\rho_1)$	0	0	0	$4(\zeta_5)$					
$2(\eta_{-1}^*)$	$16(2\rho_0)$ $\oplus$ $2(\eta\kappa_0)$	$2(\sigma_1^2)$	0	0	$8(\zeta_4)$	0	$2(\eta\epsilon_6)$				
$32(\rho_{-1})$	$2(\sigma_0^2)$	0	0	$4(\zeta_3)$	0	$2(\mu_5)$	0	$16(\sigma_7)$			
$2(\sigma_{-1}^2)$ $\oplus$ $2(\kappa_{-1})$	0	0	$8(\zeta_2)$	0	$2(\nu_4^3)$ $\oplus$ $2(\eta\epsilon_4)$	0	$8(2\sigma_6)$	0	0		
0	0	$4(\zeta_1)$	0	$2(\mu_3)$	0	$16(\sigma_5)$	0	0	0	0	
0	$8(\zeta_0)$	0	$2(\eta\epsilon_2)$	$2(\bar{\nu}_3)$	$8(2\sigma_4)$	$2(\nu_5^2)$	0	0	$2(4\nu_8)$	0	0
$4(\zeta_{-1})$	0	$2(\mu_1)$	0	$16(\sigma_3)$	0	0	0	0	0	0	$(2\iota_{10})$
0	$2(\nu_0^3)$ $\oplus$ $2(\eta\epsilon_0)$	0	$8(2\sigma_2)$	0	0	0	$2(\nu_6)$	0	0	$(4\iota_9)$	
$2(\mu_{-1})$	0	$16(\sigma_1)$	0	0	0	$2(2\nu_5)$	0	0	$(8\iota_8)$		
0	$8(2\sigma_0)$	$2(\nu_1^2)$	0	0	0	0	0	(ι_7)			
$16(\sigma_{-1})$	$2(\nu_0^2)$	0	0	0	0	0	$(4\iota_6)$				
0	0	0	0	0	0	$(4\iota_5)$					
0	0	0	0	0	$(8\iota_4)$						
0	$8(\nu_0)$	0	0	$(2\iota_3)$							
0	0	0	$(2\iota_2)$								
0	0	$(4\iota_1)$									
0	$(2\iota_0)$										
(ι_{-1})											

Table 2: E^∞ in total degrees $s+t \le 20$

0											
0	$4(2\bar{\kappa}_0)$										
$2(\bar{\kappa}_{-1})$	$2(\bar{\sigma}_0)$ $\oplus$ $8(\bar{\zeta}_0)$	0									
0	$8(\nu_0^*)$	0	0								
0	$2(\nu\kappa_0)$ $\oplus$ $2(\eta^2\rho_0)$	$2(\eta_1^*)$	$8(2\rho_2)$	$2(\sigma_3^2)$ $\oplus$ $2(\kappa_3)$							
$2(\bar{\mu}_{-1})$	0	$16(\rho_1)$	0	0	0						
0	$16(2\rho_0)$ $\oplus$ $2(\eta\kappa_0)$	$2(\sigma_1^2)$	0	0	$4(\zeta_4)$	0					
$2(\rho_{-1})$	$2(\sigma_0^2)$	0	0	0	0	0	0				
0	0	0	$2(\zeta_2)$	0	$2(\nu_4^3)$ $\oplus$ $2(\eta\epsilon_4)$	0	0	0			
0	0	$2(\zeta_1)$	0	0	0	0	0	0	0		
0	$8(\zeta_0)$	0	$2(\eta\epsilon_2)$	$2(\bar{\nu}_3)$	$4(2\sigma_4)$	$2(\nu_5^2)$	0	0	$2(4\nu_8)$	0	
0	0	0	0	0	0	0	0	0	0	0	$(2^8\iota_{10})$
0	$2(\nu_0^3)$ $\oplus$ $2(\eta\epsilon_0)$	0	$4(2\sigma_2)$	0	0	0	0	0	0	$(2^9\iota_9)$	
$2(\mu_{-1})$	0	$8(\sigma_1)$	0	0	0	0	0	0	$(2^7\iota_8)$		
0	$8(2\sigma_0)$	$2(\nu_1^2)$	0	0	0	0	0	$(2^8\iota_7)$			
$2(\sigma_{-1})$	$2(\nu_0^2)$	0	0	0	0	0	$(16\iota_6)$				
0	0	0	0	0	0	$(32\iota_5)$					
0	0	0	0	0	$(8\iota_4)$						
0	$8(\nu_0)$	0	0	$(16\iota_3)$							
0	0	0	$(2\iota_2)$								
0	0	$(4\iota_1)$									
0	$(2\iota_0)$										
(ι_{-1})											

TWO-PRIMARY HOMOTOPY OF $\mathbb{C}P_{-1}^{\infty}$

In this chapter we study the 2-primary homotopy type of the Thom spectrum $\mathbb{C}P_{-1}^{\infty}$ of minus the canonical complex line bundle over $\mathbb{C}P_{-1}^{\infty}$. We first use a reindexed Atiyah–Hirzebruch spectral sequence for stable homotopy to compute the 2-completed homotopy groups $\pi_*\left(\mathbb{C}P_{-1}^{\infty}\right)_2^{\wedge}$ in dimensions $* \leq 20$, and next compare with the Adams spectral sequence with the same abutment to determine the differentials in the latter spectral sequence in the same range of dimensions.

Only the simple cohomology calculations in Proposition 2.13 and Lemma 2.14 are needed for 3 and 4, including Theorem 4.5. The harder homotopy calculations in this section will first be applied in Section 5and onwards. The reader who is primarily interested in the spectrum cohomology of A(∗), rather than its homotopy groups, can therefore read Definition 2.12, Proposition 2.13, Lemma 2.14 and then skip ahead to Section 3.

The reindexed Atiyah–Hirzebruch spectral sequence in question is derived from the stable homotopy exact couple associated to the filtration of $\mathbb{C}P_{-1}^{\infty}$ by the subspectra $\mathbb{C}P_{-1}^{s}$, for $s \geq -1$. Its E^1-term is

$$E_{s,t}^1 = \pi_{s+t}(\mathbb{C}P_{-1}^s/\mathbb{C}P_{-1}^{s-1}) \cong \pi_{t-s}^S$$

(2.1)

for $s \geq -1$, and zero elsewhere. Here $\pi_S^k = \pi k(\Sigma^{\infty}S^0)$ is the kth stable stem.

To determine the differentials in the reindexed Atiyah–Hirzebruch spectral sequence, we compare with the computation by Mosher [34] of the differentials in the corresponding spectral sequence for the stable homotopy of $\mathbb{C}P^{\infty}$. The E^1-term of the latter spectral sequence is obtained from (2.1) by restricting to filtrations $s \geq 1$, i.e., by omitting the columns s=−1 and 0, and the collapse map $j: \mathbb{C}P_{-1}^{\infty} \to \mathbb{C}P^{\infty}$ induces a map of spectral sequences. From here on we often use the same notation for a based space and its suspension spectrum, such as writing S^0 for $\Sigma^\infty S^0$.

The differentials in (2.1) landing in filtration s=0 are always zero, due to the splitting $\mathbb{CP}_0^\infty = \mathbb{CP}_+^\infty \simeq \mathbb{CP}^\infty \vee S^0$. The differentials in (2.1) landing in filtration s=−1 arise from the connecting map in the cofiber sequence $S^{-2} \to \mathbb{CP}_{-1}^\infty \to \mathbb{CP}_+^\infty$. This is the wedge sum of the (desuspended) S^1-transfer map $\mathbb{CP}^\infty \to S^{-1}$, and the (desuspended) multiplication by η map $S^0 \to S^{-1}$. The image of the S^1-transfer map was computed in dimensions $*\leq 20$ by Mukai in [35], [36] and [37], and we use these results to determine the differentials in (2.1) landing in filtration s=−1 in the same range of dimensions.

For ease of reference we use similar notation for classes in our spectral sequence (2.1) as in [34]. Thus we write $\beta_s \in E^1_{s,s+1}$ for the class corresponding to $\beta \in E^s_t$, and write $\mathbb{Z}/n(\beta)$ for a cyclic group of order n with generator β. In Table 1 and Table 2 we briefly write $n(\beta)$ for $\mathbb{Z}/n(\beta)$ and (β) for $\mathbb{Z}/(\beta)$, to save some space. Hereafter we concentrate on the 2-primary components, and all spectra and groups are implicitly 2-completed. Differentials are mostly given only up to multiplication by a 2-adic unit.

In dimensions $*\leq 22$, we will use the following presentation for the stable stems π_*^S, following the tables in [50, XIV; 40, A3.3].

$$\pi_0^S = \mathbb{Z}(\iota), \ \pi_1^S = \mathbb{Z}/2(\eta), \ \pi_2^S = \mathbb{Z}/2(\eta^2), \ \pi_3^S = \mathbb{Z}/8(\nu), \ \pi_4^S = 0, \ \pi_5^S = 0, \ \pi_6^S = \mathbb{Z}/2(\nu^2), \ \pi_7^S = \mathbb{Z}/16(\sigma),$$
$$\pi_8^S = \mathbb{Z}/2(\bar{\nu}) \oplus \mathbb{Z}/2(\varepsilon), \ \pi_9^S = \mathbb{Z}/2(\nu^3) \oplus \mathbb{Z}/2(\eta\varepsilon) \oplus \mathbb{Z}/2(\mu), \ \pi_{10}^S = \mathbb{Z}/2(\eta\mu), \ \pi_{11}^S = \mathbb{Z}/8(\zeta), \ \pi_{12}^S = 0, \ \pi_{13}^S = 0,$$
$$\pi_{14}^S = \mathbb{Z}/2(\sigma^2) \oplus \mathbb{Z}/2(\kappa), \ \pi_{15}^S = \mathbb{Z}/32(\rho) \oplus \mathbb{Z}/2(\eta\kappa), \ \pi_{16}^S = \mathbb{Z}/2(\eta^*) \oplus \mathbb{Z}/2(\eta\rho), \ \pi_{17}^S = \mathbb{Z}/2(\eta\eta^*) \oplus$$
$$\mathbb{Z}/2(\nu\kappa) \oplus \mathbb{Z}/2(\eta^2\rho) \oplus \mathbb{Z}/2(\bar{\mu}), \ \pi_{18}^S = \mathbb{Z}/8(\nu^*) \oplus \mathbb{Z}/2(\eta\bar{\mu}), \ \pi_{19}^S = \mathbb{Z}/2(\bar{\sigma}) \oplus \mathbb{Z}/8(\bar{\zeta}), \ \pi_{20}^S = \mathbb{Z}/8(\bar{\kappa}),$$
$$\pi_{21}^S = \mathbb{Z}/2(\nu\nu^*) \oplus \mathbb{Z}/2(\eta\bar{\kappa}) \text{ and } \pi_{22}^S = \mathbb{Z}/2(\nu\bar{\sigma}) \oplus \mathbb{Z}/2(\eta^2\bar{\kappa}).$$

For a fixed r, the dr-differentials in the spectral sequence for $\pi_*^S(\mathbb{CP}^\infty)$ are periodic in the filtration degree s, see [34, 4.4], and this periodicity propagates to the spectral sequence (2.1). Hence Mosher's description of the d^1-, d^2- and d^3-differentials for $\mathbb{CP}^\infty$ in [34, 5.1, 5.2, and 5.4] extends to give Proposition 2.2, Proposition 2.3 and Proposition 2.4 for the corresponding differentials in (2.1). Let $\beta \in \pi_*^S$.

Proposition 2.2

$d^1(\beta_s)=0$ for s odd and $d^1(\beta_s)=\eta\beta_{s-1}$ for s even.

Proposition 2.3

$d^2(\beta_s)=\nu\beta_{s-2}$ for $s \equiv 0,1,4,5 \bmod 8$, $d^2(\beta_s)=2\nu\beta_{s-2}$ for $s \equiv 3,6 \bmod 8$ and $d^2(\beta_s)=0$ for $s \equiv 2,7 \bmod 8$.

Proposition 2.4

$d^3(\beta_s)=0$ for s odd. If s is even then $d^3(\beta_s)=\gamma_{s-3}$, where $\gamma \in \langle \eta,\nu,\beta \rangle$ for $s \equiv 0 \bmod 8$ $\gamma \in \langle \nu,\eta,\beta \rangle$ for $s \equiv 2 \bmod 8$ $\gamma \in \langle 2\nu,\eta,\beta \rangle + \langle \eta,\nu,\beta \rangle$ for $s \equiv 4 \bmod 8$ and $\gamma \in \langle \nu,\eta,\beta \rangle + \langle \eta,2\nu,\beta \rangle$ for $s \equiv 6 \bmod 8$.

The d^1-differentials in (2.1) are given by the following multiplicative relations in π_*^S, see [40] and [50].

$$\eta \cdot \iota = \eta, \ \eta \cdot \eta = \eta^2, \ \eta \cdot \eta^2 = 4\nu, \ \eta \cdot \nu = 0, \ \eta \cdot \nu^2 = 0, \ \eta \cdot \sigma = \bar{\nu} + \varepsilon, \ \eta \cdot \bar{\nu} = \nu^3, \ \eta \cdot \varepsilon = \eta\varepsilon, \ \eta \cdot \nu^3 = 0,$$
$$\eta \cdot \eta\varepsilon = 0, \ \eta \cdot \mu = \eta\mu, \ \eta \cdot \eta\mu = 4\zeta, \ \eta \cdot \zeta = 0, \ \eta \cdot \sigma^2 = 0, \ \eta \cdot \kappa = \eta\kappa, \ \eta \cdot \rho = \eta\rho, \ \eta \cdot \eta\kappa = 0, \ \eta \cdot \eta^* = \eta\eta^*,$$
$$\eta \cdot \eta\rho = \eta^2\rho, \ \eta \cdot \eta\eta^* = 4\nu^*, \ \eta \cdot \nu\kappa = 0, \ \eta \cdot \eta^2\rho = 0, \ \eta \cdot \bar{\mu} = \eta\bar{\mu}, \ \eta \cdot \nu^* = 0, \ \eta \cdot \eta\bar{\mu} = 4\bar{\zeta}, \ \eta \cdot \bar{\sigma} = 0,$$
$$\eta \cdot \bar{\zeta} = 0, \ \eta \cdot \bar{\kappa} = \eta\bar{\kappa}, \ \eta \cdot \nu\nu^* = 0 \text{ and } \eta \cdot \eta\bar{\kappa} = \eta^2\bar{\kappa}.$$

For example, $\bar{\sigma} = \langle \nu,\eta\sigma,\sigma \rangle$, so $\eta.\bar{\sigma} = -\langle \nu,\eta\sigma,\sigma \rangle\sigma \simeq 0$ with zero indeterminacy.

The d^2-differentials in (2.1) are given by the following multiplicative relations in π_*^S, see [40] and [50].

$$\nu \cdot \iota = \nu, \ \nu \cdot \nu = \nu^2, \ \nu \cdot \nu^2 = \nu^3, \ \nu \cdot \sigma = 0, \ \nu \cdot \bar{\nu} = 0, \ \nu \cdot \nu^3 = 0, \ \nu \cdot \eta\varepsilon = 0, \ \nu \cdot \mu = 0, \ \nu \cdot \zeta = 0, \ \nu \cdot \sigma^2 = 0,$$
$$\nu \cdot \kappa = \nu\kappa, \ \nu \cdot \rho = 0, \ \nu \cdot \eta\kappa = 0, \ \nu \cdot \eta^* = 0, \ \nu \cdot \nu\kappa = 4\bar{\kappa}, \ \nu \cdot \eta^2\rho = 0, \ \nu \cdot \bar{\mu} = 0, \ \nu \cdot \nu^* = \nu\nu^*, \ \nu \cdot \eta\bar{\mu} = 0,$$
$$\nu \cdot \bar{\sigma} = \nu\bar{\sigma} \text{ and } \nu \cdot \bar{\zeta} = 0.$$

The d^3-differentials are given by the following secondary compositions, from [33; 34, 10.1; 50].

$$\langle v, \eta, v \rangle = \bar{v}, \ \langle \eta, v, 2v \rangle = \langle \eta, 2v, v \rangle = \{\varepsilon, \bar{v}\}, \ \langle v, \eta, \zeta \rangle \subseteq \{0, \eta\rho\}, \ \langle \eta, v, \zeta \rangle = \{0, \eta\rho\}, \ \langle v, \eta, \sigma^2 \rangle = \bar{\sigma},$$

$\langle v, \eta, 2\rho \rangle = \{0, 4\bar{k}\}$ and $\eta.\bar{\sigma} = -\langle v, \eta, \eta k \rangle = \pm 2\bar{k}$ by Mimura and Toda [33].

The resulting E^4-term is shown in Table 1, accounting for all differentials landing in total degree $s+t \leq 20$.

In Lemmas 2.5–2.10, we only consider differentials landing in total degree $s+t \leq 20$.

Lemma 2.5

The non-zero d^4-differentials in (2.1) are $d^4(2\iota_3)=2\sigma_{-1}$, $d^4(4\iota_5)=8\sigma_1$, $d^4(4\iota_6)=8\sigma_2$, $d^4(\iota_7)=2\sigma_3$, $d^4(8\iota_8)=8\sigma_4$, $d^4(4\iota_9)=4\sigma_5$, $d^4(2\iota_{10})=2\sigma_6$ and $d^4(\sigma_3)=\sigma_{-1}^2$.

Proof

The d^4-differentials landing in filtration $s \geq 1$ and total degree $s+t \leq 19$ are determined by those in the spectral sequence for $\pi_*^S(\mathbb{CP}^\infty)$, and are given in [34, 5.6 and 6.4].

In total degree 20, $d^4(\zeta_5)=0$ by the computation of $\pi_{20}^S(\mathbb{CP}^\infty)$ following [37, 4.2], and $d^4(\sigma_7)=0$ by the proof of [37, 4.3] (the formula $\gamma_6\sigma = 2\bar{i}\bar{\sigma}'\sigma$).

The differentials landing in filtration $s=0$ are always zero, as noted above. The differentials landing in filtration $s=-1$ are determined by the computation of the S^1-transfer in [35] and [36]. Thus $d^4(2\iota_3)=2\sigma_{-1}$ by Mukai [35, 13.1(iii)], $d^4(\sigma_3)=\sigma_{-1}^2$ by the proof of [36, 5.3] (the formula $g_4\tilde{\sigma}'=\sigma^2$), $d^4(\bar{v}_3)$ by the proof of [36, 5.3] (the formula $g_8\bar{i}\bar{v}=\eta k$),

$d^4(\mu_3)=0$ by the proof of [36, 5.4] (the formula $g_4\tilde{\mu}=0$), and $d^4(\zeta_3)=0$ by the proof of [36, 5.5] (the formula $g_4\pi^S_{17}(\mathbb{CP}^3)=0$).

Lemma 2.6

The non-zero d^5-differentials in (2.1) are $d^5(8\iota_6)=\mu_1$, $d^5(16\iota_8)=\mu_3$ and $d^5(16\iota_{10})=\mu_5$.

Proof

The d^5-differentials landing in filtration $s\geq 1$ and total degree $s+t\leq 19$ are determined by those in the spectral sequence for $\pi^S_*(\mathbb{CP}^\infty)$, and are given in [34, 6.5].

In total degree 20, $d^5(\eta\varepsilon_6)=0$ by the calculation of $\pi^S_{20}(\mathbb{CP}^6)$ following [37, 4.2].

The differentials landing in filtration $s=-1$ are $d^5(8\iota_4)=0$ by Mukai [35, 13.1(iv)], $d^5(2\sigma_4)=0$ by the proof of [36, 5.4] (the formulas $g_5\widetilde{2\sigma'}\equiv 0 \bmod \mu\sigma$), $d^5(v_4^3)=0$ and $d^5(\eta\varepsilon_4)=0$ by the proof of [36, 5.5] (the formulas $g_5\tilde{v}^3\equiv 0 \bmod\{4v^*, \eta\bar{\mu}\}$ and $g_5\lambda=0$, where λ was chosen as a coextension of $\eta^2\sigma$ before [36, 4.7]), and $d^5(\zeta_4)=0$ by Mukai [37, 5.1].

Lemma 2.7

The non-zero d^6-differentials in (2.1) are $d^6(8\iota_5)=\zeta_{-1}$, $d^6(8\iota_7)=2\zeta_1$, $d^6(32\iota_8)=2\zeta_2$, $d^6(16\iota_9)=\zeta_3$, $d^6(32\iota_{10})=4\zeta_4$, $d^6(2v_5)=\kappa_{-1}$, $d^6(\sigma_5)=v^*_{-1}$ and $d^6(\sigma_7)=2v^*_1$.

Proof

The differentials landing in filtration $s\geq 1$ and total degree $s+t\leq 19$ come from [34, 6.6], and $d^6(\sigma_7)=2v^*_1$ by Mukai [37, 4.3] and its proof (namely, $\gamma_6\sigma=2\widetilde{1\tilde{\sigma}}\sigma=2iv^*$.

Also $d^6(8\iota_5) = \zeta_{-1}$ by Mukai [35, 13.1(v)], $d^6(2v_5) = \kappa_{-1}$ by the proof of [36, 5.3] (the formula $g_8\left(\widetilde{i2v}''\right) = k \bmod \sigma^2$), $d^6(v_5{}^2) = 0$ by the proof of [36, 5.4] (the formula $g_0 i\tilde{v}^2 \equiv \omega^*\eta \bmod ivk$), and $d^6\left(\sigma_5\right) = \pm v^*_{-1}$ by the proof of [36, 5.5] (the formula $g_6\tilde{\sigma}'' \equiv xv^*\eta \bmod \eta\bar{\mu}$ where x is odd).

Lemma 2.8

The only non-zero d^7-differential in (2.1) is $d^7\left(\sigma_6\right) = \eta^*_{-1}$.

Proof

We have $d^7(\sigma_6) = \eta^*_{-1}$ by Mukai [36, 5.4] and its proof (the formula $g_7\tilde{v}'' \equiv \omega^*$). All other d^7-differentials are zero by Mosher [34, 6.7] or bidegree reasons.

Lemma 2.9

The non-zero d^8-differentials in (2.1) are $d^8(16\iota_7) = 2\rho_{-1}$, $d^8(64\iota_9) = 16\rho_1$, and $d^8(64\iota_{10}) = 16\rho_2$.

Proof

These follow from [35, 4.3] since 2ρ generates the complex image of J in dimension 15, and from [34, 6.8].

Lemma 2.10

The remaining non-zero differentials in (2.1) are $d^9\left(2^7\iota_{10}\right) = \bar{\mu}_1$ and $d^{10}\left(2^7\iota_9\right) = \bar{\zeta}_1$.

Proof

These follow from [34, 6.9; 35, 4.3], since $\bar{\zeta}$ generates the complex image of J in dimension 19.

This leaves us with the E^∞-term shown in Table 2, in total degrees $s+t \leq 20$. Recall the convention that $n(\beta)$ denotes a cyclic group of order n, generated by β.

Theorem 2.11

The 2-primary homotopy groups of $\mathbb{CP}^\infty_{-1}$ in dimensions $* \leq 20$ are as follows:

$$\pi_{-2}(\mathbb{CP}^\infty_{-1}) = \mathbb{Z}(\iota_{-1}),$$

$$\pi_{-1}(\mathbb{CP}^\infty_{-1}) = 0,$$

$$\pi_0(\mathbb{CP}^\infty_{-1}) = \mathbb{Z}(2\iota_0),$$

$$\pi_1(\mathbb{CP}^\infty_{-1}) = 0,$$

$$\pi_2(\mathbb{CP}^\infty_{-1}) = \mathbb{Z}(4\iota_1),$$

$$\pi_3(\mathbb{CP}^\infty_{-1}) = \mathbb{Z}/8(\nu_0),$$

$$\pi_4(\mathbb{CP}^\infty_{-1}) = \mathbb{Z}(2\iota_2),$$

$$\pi_5(\mathbb{CP}^\infty_{-1}) = \mathbb{Z}/2(\sigma_{-1}),$$

$$\pi_6(\mathbb{CP}^\infty_{-1}) = \mathbb{Z}/2(\nu_0^2) \oplus \mathbb{Z}(16\iota_3),$$

$$\pi_7(\mathbb{CP}^\infty_{-1}) = \mathbb{Z}/2(\mu_{-1}) \rtimes \mathbb{Z}/8(2\sigma_0)$$

$$\cong \mathbb{Z}/16(2\sigma_0),$$

$$\pi_8(\mathbb{CP}^\infty_{-1}) = \mathbb{Z}/2(\nu_1^2) \oplus \mathbb{Z}(8\iota_4),$$

$$\pi_9(\mathbb{C}P^\infty_{-1}) = \mathbb{Z}/2(\nu_0^3) \oplus \mathbb{Z}/2(\eta\varepsilon_0) \oplus \mathbb{Z}/8(\sigma_1),$$

$$\pi_{10}(\mathbb{C}P^\infty_{-1}) = \mathbb{Z}(32\iota_5),$$

$$\pi_{11}(\mathbb{C}P^\infty_{-1}) = \mathbb{Z}/8(\zeta_0) \oplus \mathbb{Z}/4(2\sigma_2),$$

$$\pi_{12}(\mathbb{C}P^\infty_{-1}) = \mathbb{Z}(16\iota_6),$$

$$\pi_{13}(\mathbb{C}P^\infty_{-1}) = \mathbb{Z}/2(\rho_{-1}) \rtimes \mathbb{Z}/2(\zeta_1) \rtimes \mathbb{Z}/2(\eta\varepsilon_2)$$

$$\cong \mathbb{Z}/2(\rho_{-1}) \rtimes \mathbb{Z}/4(\eta\varepsilon_2),$$

$$\pi_{14}(\mathbb{C}P^\infty_{-1}) = \mathbb{Z}/2(\sigma_0^2) \oplus \mathbb{Z}/2(\bar{\nu}_3) \oplus \mathbb{Z}(2^8\iota_7),$$

$$\pi_{15}(\mathbb{C}P^\infty_{-1}) = \mathbb{Z}/2(\bar{\mu}_{-1}) \rtimes \mathbb{Z}/16(2\rho_0) \oplus \mathbb{Z}/2(\eta\kappa_0) \rtimes \mathbb{Z}/2(\zeta_2) \rtimes \mathbb{Z}/4(2\sigma_4)$$

$$\cong \mathbb{Z}/32(2\rho_0) \oplus \mathbb{Z}/2(\eta\kappa_0) \rtimes \mathbb{Z}/2(\zeta_2) \rtimes \mathbb{Z}/4(2\sigma_4),$$

$$\pi_{16}(\mathbb{C}P^\infty_{-1}) = \mathbb{Z}/2(\sigma_1^2) \oplus \mathbb{Z}/2(\nu_5^2) \oplus \mathbb{Z}(2^7\iota_8),$$

$$\pi_{17}(\mathbb{C}P^\infty_{-1}) = \mathbb{Z}/2(\nu\kappa_0) \oplus \mathbb{Z}/2(\eta^2\rho_0) \oplus \mathbb{Z}/16(\rho_1) \rtimes \mathbb{Z}/2(\nu_4^3) \oplus \mathbb{Z}/2(\eta\varepsilon_4)$$

$$\cong \mathbb{Z}/2(\nu\kappa_0) \oplus \mathbb{Z}/2(\eta^2\rho_0) \oplus \mathbb{Z}/32(\eta\varepsilon_4) \oplus \mathbb{Z}/2(\nu_4^3),$$

$$\pi_{18}(\mathbb{C}P^\infty_{-1}) = \mathbb{Z}/2(\bar{\kappa}_{-1}) \rtimes \mathbb{Z}/8(\nu_0^*) \rtimes \mathbb{Z}/2(\eta_1^*) \oplus \mathbb{Z}(2^9\iota_9),$$

$$\pi_{19}(\mathbb{C}P^\infty_{-1}) = \mathbb{Z}/2(\bar{\sigma}_0) \oplus \mathbb{Z}/8(\bar{\zeta}_0) \oplus \mathbb{Z}/8(2\rho_2) \rtimes \mathbb{Z}/4(\zeta_4) \rtimes \mathbb{Z}/2(4\nu_8)$$

$$\cong \mathbb{Z}/2(\bar{\sigma}_0) \oplus \mathbb{Z}/8(\bar{\zeta}_0) \oplus \mathbb{Z}/64(4\nu_8),$$

$$\pi_{20}(\mathbb{C}P^\infty_{-1}) = \mathbb{Z}/4(2\bar{\kappa}_0) \rtimes \mathbb{Z}/2(\sigma_3^2) \oplus \mathbb{Z}/2(\kappa_3) \oplus \mathbb{Z}(2^8\iota_{10}).$$

Proof

Up to extensions, this can be read off from the E^∞-term above.

In dimensions $*=9,11,14,17,19$ the subgroup in filtration s=0 is split off by the composite map $\mathbb{CP}^\infty_{-1} \to \mathbb{CP}^\infty_+ \to S^0$, followed by a retraction of π^S_* onto the kernel of $\eta : \pi^S_* \to \pi^S_{*+1}$.

The extension in dimension 7 will follow from the proof of Proposition 2.18 below, in view of h_0-multiplications in the Adams spectral sequence for $\pi_*(\mathbb{CP}^\infty_{-1})$.

The right hand extension in dimension 13 can be read off from

$$\pi^S_{13}(\mathbb{C}P^2) \cong \mathbb{Z}/8(\eta\varepsilon_2) \oplus \mathbb{Z}/2(v_2^3),$$

See [36, p. 197].

The left hand extension in dimension 15 can be read off from $\pi^S_{19}(\mathbb{C}P^2) \cong \pi^S_{15}(\mathbb{C}P^0_{-1})$, see [37, p. 133].

The splitting in dimension 16 can be deduced from the injection

$$\pi_{16}(\mathbb{C}P^\infty_{-1}) \to \pi_{16}(\mathbb{C}P^\infty) \cong (\mathbb{Z}/2)^3 \oplus \mathbb{Z}, \text{ see } [36, 1(ii)].$$

The right hand extension in dimension 17 can be read off from $\pi^S_{17}(\mathbb{C}P^4)$, see [36, 4.7 and 4.8]. Note that $\eta^2\sigma=v^3+\eta\varepsilon$, so twice the coextension λ of $\eta^2\sigma$ is twice a coextension of $\eta\varepsilon$.

The middle and right hand extensions in dimension 19 follow from [37, 3.2].

We proceed to compare these results with the Adams spectral sequence for $\pi^S_*(\mathbb{C}P^\infty_{-1})^\wedge_2$. Let $A = A(2)$ be the mod 2 Steenrod algebra, generated by the Steenrod squaring operations Sq^i. For each sequence of natural numbers $I = (i_1,...,i_n)$ let $Sq^I = Sq^{i_1} \circ \cdots \circ Sq^{i_n}$ be the composite operation.

The sequence I, or the operation Sq^I, is said to be admissible if $i_s \geq 2i_{s+1}$ for all $0 \leq s < n$. The set of admissible Sq^I form a vector space basis for A.

Definition 2.12

Let C be the left ideal in A with vector space basis the set of admissible Sq^I such that $I = (i_1, \ldots, i_n)$ has length $n \geq 2$, or $I = (i)$ with i odd. Then A/C is a cyclic left A-module, with vector space basis the set of Sq^i with i ≥ 0 even.

Let us briefly write $H^*(X)$ for the mod 2 spectrum cohomology $H^*_{Spec}(X; F_2)$ of a spectrum X. It is naturally a graded left A-module.

Proposition 2.13

$$H^*(\mathbb{CP}^\infty_{-1}) \cong \Sigma^{-2} A/C$$

as graded leftA-modules.

Proof

It is clear that $H^n(\mathbb{CP}^\infty_{-1}) \cong F_2$ for $n \geq -2$ even, and 0 otherwise. In $H^*(\Sigma^\infty \mathbb{CP}^\infty_+) \cong F_2\{y^j \mid j \geq 0\}$ with deg(y) $=2$ the squaring operations are given by $Sq^{2i-1}(y^j)=0$ and $Sq^{2i}(y^j)=\binom{j}{i}y^{i+j}$. By James periodicity and stability of the squaring operations the same formulas apply in

$$H^*(\mathbb{CP}^\infty_{-1}) \cong F_2\{y^j \mid j \geq -1\},$$

Including the case j$=-1$. Then $Sq^{2i}(y^{-1})=y^{i-1}$ since $\binom{-1}{i} \equiv 1 \bmod 2$. To prove the proposition it remains to show that $Sq^I(y^{-1})=0$ when $I=(i_1, \ldots, i_n)$ is admissible of length ≥ 2. Let $z=Sq^{i_n}(y^{-1})$. Then z has dimension (i_n-2) and lifts to the ordinary cohomology $H^*(\mathbb{CP}^\infty_+; F_2)$ of the space $\mathbb{CP}^\infty_+$, which is an unstable A-module. Thus $Sq^{i_{n-1}}(z) = 0$ since $i_{n-1} > i_n - 2$, and so $Sq^i(y^{-1}) = 0$.

Lemma 2.14

In $\mathbb{C}P^\infty_{-1}$ the lowest k-invariant

$$k^1 : \Sigma^{-2} H\mathbb{Z} \to \Sigma H\mathbb{Z}$$

is non-trivial, and has mod 2 reduction the class of Sq^3 mod ASq^1.

Proof

The lowest homotopy group of $\mathbb{C}P^\infty_{-1}$ is detected by a map $\mathbb{C}P^\infty_{-1} \to \Sigma^{-2} H\mathbb{Z}$. On cohomology it induces a surjection $\Sigma^{-2}A/ASq^1 \to \Sigma^{-2}A/C$, whose kernel $\Sigma^{-2}C/ASq^1$ begins with $\Sigma^{-2} Sq^3$ mod ASq^{-1} in degree 1. This is the cohomology operation represented by the lowest k-invariant k^1.

Consider the Adams spectral sequence

$$E_2^{s,t} = \mathrm{Ext}_A^{s,t}(H^*(\mathbb{C}P^\infty_{-1}), \mathbb{F}_2) \Rightarrow \pi_{t-s}(\mathbb{C}P^\infty_{-1})_2^\wedge.$$

$$(2.15)$$

Its E_2-term can be computed in a range from a (minimal) resolution of $\Sigma^{-2}A/C$, either by hand or by Bruner's Ext-calculator program [12]. The E_2-term in homotopical degrees $t-s \leq 20$ is shown in Table 3a and Table 3b. The notation $_s x$ represents a class arising in the Adams E_2-term for $\mathbb{C}P^\infty_{-1}$, mapping to the class usually named x in the Adams E_2-term for $= \mathbb{C}P^s_{-1} / \mathbb{C}P^{s-1}_{-1} \cong \Sigma^{-2s} S^0$. The distinction between classes marked as '•' or as '∘' will be explained in Section 5, following the long exact sequence (5.5). Classes supporting or hit by differentials are shaded gray. In bidegree $(3, 14)$ a d_2-differential hits the sum of the two given generators, which are therefore both half-shaded.

The cofiber sequence of spectra

$$\mathbb{C}P^0_{-1} \xrightarrow{i} \mathbb{C}P^\infty_{-1} \xrightarrow{j} \mathbb{C}P^\infty$$

Induces a short exact sequence in mod 2 spectrum cohomology, and thus gives a long exact sequence of Ext-groups relating the Adams E_2-term (2.15) to the Adams E_2-terms

$$'E_2^{s,t} = \mathrm{Ext}_A^{s,t}(H^*(\mathbb{C}P^0_{-1}), \mathbb{F}_2) \Rightarrow \pi_{t-s}(\mathbb{C}P^0_{-1})_2^\wedge \qquad (2.16)$$

And

$$''E_2^{s,t} = \mathrm{Ext}_A^{s,t}(H^*(\mathbb{C}P^\infty), \mathbb{F}_2) \Rightarrow \pi_{t-s}(\mathbb{C}P^\infty)_2^\wedge. \qquad (2.17)$$

Knowledge of the stable homotopy of $\mathbb{C}P^0_{-1} \simeq \Sigma^{-4}\mathbb{C}P^2$ and $\mathbb{C}P^\infty$ in a range allows us to determine the differentials in the spectral sequences $'E_*$ and $''E_*$ in a similar range. This is comparatively easy for $\mathbb{C}P^0_{-1}$, and was done for $\mathbb{C}P^\infty$ by Mosher [34]. Using the long exact sequence of E_2-terms

$$\cdots \longrightarrow {'E_2^{s,t}} \xrightarrow{i_*} E_2^{s,t} \xrightarrow{j_*} {''E_2^{s,t}} \xrightarrow{\partial} {'E_2^{s+1,t}} \longrightarrow \cdots$$

And the geometric boundary theorem [40, 2.3.4] we can transfer some of these differentials to (2.15). (The careful reader should come equipped with the Ext charts for $\mathbb{C}P^0_{-1}$ and $\mathbb{C}P^\infty$ to check the details in the following proof.)

Proposition 2.18

In the Adams spectral sequence (2.15) the non-zero differentials landing in homotopical degree ≤ 20 are:

i. $d_2^{1,8}(_3h_1) = {}_{-1}c_0$.

ii. $d_2^{2,12}(_5h_0^2) = h_0^3 \cdot {}_1h_3$, $d_2^{3,13}(_5h_0^3) = h_0^4 \cdot {}_1h_3 = {}_{-1}Ph_2$ and $d_2^{4,14}(_5h_0^4)$
    and $d_2^{4,14}(_5h_0^4) = h_0^5 \cdot {}_1h_3 = {}_{-1}h_0Ph_2$.

iii. $d_2^{1,13}(_6h_0) = h_0 \cdot _2h_0h_3 + h_1 \cdot _5h_0^2, \; d_2^{2,14}(_6h_0^2) = h_0^2 \cdot _2h_0h_3, \; d_2^{3,15}(_6h_0^3) = h_0^3 \cdot _2h_0h_3$
and $d_2^{2,15}(_5h_0h_2) = {}_{-1}d_0.$

iv. $d_2^{1,16}(_6h_2) = h_0 \cdot _0h_3^2, \; d_3^{2,17}(h_0 \cdot _6h_2) = _0h_0d_0, \; d_3^{3,18}(h_0^2 \cdot _6h_2) = h_0 \cdot _0h_0d_0$
and $d_3^{4,19}(h_0^3 \cdot _6h_2) = h_0^2 \cdot _0h_0d_0 = {}_{-1}Pc_0.$

v. $d_2^{1,18}(_5h_3) = h_0 \cdot _1h_3^2, \; d_2^{2,19}(h_0 \cdot _5h_3) = h_0^2 \cdot _1h_3^2 = {}_{-1}f_0$ and
$d_3^{3,20}(h_0^2 \cdot _5h_3) = _1h_0^2d_0.$

vi. $d_2^{4,22}(x) = h_0^2 \cdot _4h_1c_0$ with $h_0 \cdot x \neq 0, \; d_2^{5,23}(h_0 \cdot x) = h_0^3 \cdot _4h_1c_0, \; d_2^{6,24}(_9h_0^6) = h_0^7 \cdot _5h_3, \; d_2^{7,25}(h_0 \cdot _9h_0^6) = h_0^8 \cdot _5h_3 = {}_{-1}P^2h_2$ and $d_2^{8,26}(h_0^2 \cdot _9h_0^6) = h_0^9 \cdot _5h_3 = {}_{-1}h_0P^2h_2.$

vii. $d_2^{4,24} \neq 0, \; d_2^{5,25} \neq 0, \; d_2^{6,26} \neq 0, \; d_2^{7,27} \neq 0, \; d_2^{1,22}(_3h_4) \neq 0$ and $d_2^{6,27} \neq 0$ all have rank 1.

Proof

We compare the Adams E_2-term in Table 3 with its abutment (Theorem 2.11). Each h_0-torsion class in the E_∞-term of (2.15) comes from an h_0-torsion class in the E_2-term, and so is represented by a 2-torsion class in $\pi_*(\mathbb{CP}_{-1}^\infty)$. (The proof of this assertion goes by induction over the subspectra $\mathbb{CP}_{-1}^s$ of $\mathbb{CP}_{-1}^\infty$.)

In each degree $t-s \leq 5$ the order of the 2-torsion in $\pi_*(\mathbb{CP}_{-1}^\infty)$ equals the order of the h_0-torsion in Table 3, hence there are no non-zero differentials in this range.

i.   In degree $t-s=6$ the 2-torsion in the abutment is $\mathbb{Z}/2$, while the E_2-term has two h_0-torsion generators, so one of these must be hit by a differential. For bidegree reasons the only possibility is $d_2^{1,8}(_3h_1) = {}_{-1}c_0$, and then there is no room for further differentials landing in degrees $t-s \leq 8$.

In degree 7 of the E_∞-term there is then a non-zero multiplication by h_0^3, showing that the extension in $\pi_7\left(\mathbb{CP}_{-1}^\infty\right)$ is cyclic.

ii. and

iii. (iii) We turn to degrees $9 \leq t-s \leq 13$. The Adams spectral sequence for $\mathbb{CP}^\infty$, denoted $''E_*$ in (2.17), has differentials $''d_2\left({}_5h_0^2\right) =_1 h_0^3 h_3$ and $''d_2\left({}_6h_0\right) = h_0 \cdot_2 h_0 h_3 + h_1 \cdot_5 h_0^2$. This uses $\pi_9^S(\mathbb{CP}^\infty) = \mathbb{Z}/2$ and $\pi_{11}^S(\mathbb{CP}^\infty) = \mathbb{Z}/4$, see [34, 7.2].

The map of spectral sequences $j_* : E_2 \to {}''E_2$ is an isomorphism in bidegrees $(2,12)$ and $(1,13)$, so these differentials lift to E_2.

Regarding the first $''d_2$-differential, both basis elements in $E_2^{4,13} \cong F_2\{h_0^3 \cdot_1 h_3, h_1, h_1 \cdot_4 h_0^3\}$ map to $_1 h_0^3 h_3$ in $''E_2^{4,13}$. Hence $d_2({}_5h_0^2)$ equals one or the other of these basis elements. It cannot be $h_1 \cdot_4 h_0^3$, because then $d_2({}_5h_0^3)=0$ by h_0-multiplication, and more classes would survive to the E_∞-term in degree 9 than the abutment $\pi_9\left(\mathbb{CP}_{-1}^\infty\right) \cong \mathbb{Z}/2 \oplus \mathbb{Z}/2 \oplus \mathbb{Z}/28$ allows. Thus $d_2^{2,12}\left({}_5h_0^2\right) = h_0^3 \cdot_1 h_3$. Multiplication by h_0 implies $d_2^{3,13}\left({}_5h_0^3\right) = h_0^4 \cdot_1 h_3$ and $d_2^{4,14}\left({}_5h_0^4\right) = h_0^5 \cdot_1 h_3$ in (2.15).

In bidegrees $(2,12),(2,13)$ and $(3,14)$ the map j_* is an iso-morphism, so the second $''d_2$-differential lifts to $d_2^{1,13}\left({}_6h_0\right) = h_0 \cdot_2 h_0 h_3 + h_1 \cdot_5 h_0^2$. Multiplication by h_0, h_0^2 and h_1 implies $d_2^{2,14}\left({}_6h_0^2\right) = h_0^2 \cdot_2 h_0 h_3$, $d_2^{3,15}\left({}_6h_0^3\right) = h_0^3 \cdot_2 h_0 h_3$ and $d_2^{2,15}\left({}_5h_0 h_2\right) =_{-1} d_0$, respectively. There is no room for further differentials landing in degree $t-s \leq 13$.

iv. We turn to degrees $14 \le t-s \le 15$. For bidegree reasons the class $_0 h_3^2 \in E_2^{2,16}$ survives to E_∞, and the classes $h_0 \cdot_0 h_3^2$ and $_3 c_0$ in $E_2^{3,17}$ can only be affected by a d_2-differential from $_6 h_2 \in E_2^{1,16}$. The 2-torsion in $\pi_{14}(\mathbb{CP}_{-1}^\infty)$ is $(\mathbb{Z}/2)^2$, so the class $h_0 \cdot_0 h_3^2$ cannot survive to E_∞, i.e., there is a non-zero differential $d_2(_6 h_2) = h_0 \cdot_0 h_3^2$ in E_*.

The Adams spectral sequence for $\mathbb{CP}_{-1}^0$, denoted $'E_*$ in (2.16), has a differential $d_3(_0 h_0 h_4) =_0 h_0 d_0$. (This lifts the usual differential $d_3(h_0 h_4) = h_0 d_0$ in the Adams spectral sequence for π_*^S. Multiplying this by h_0^2 gives the differential $d_3(_0 h_0^3 h_4) =_{-1} Pc_0$, arising from the hidden multiplicative relation $\eta \cdot \{h_0^3 h_4\} = \{Pc_0\}$ in the stable 16-stem.)

The map of spectral sequences $i_* : {'E_2} \to E_2$ is injective in bidegree (2, 17), taking $_0 h_0 h_4$ to $h_0 \cdot_6 h_2$. For in $'E_2$ we know that $h_2 \cdot_0 h_0 h_4 = h_0 \cdot h_2 h_4$. Thus the image of $_0 h_0 h_4$ in E_2 is such that h_2 times it is divisible by h_0, and by inspection this property characterizes $h_0 \cdot 6 h_4 \in E_2^{2,17}$. Thus we have another non-zero differential $d_3(h_0 \cdot_6 h_2) = _0 h_0 d_0$ in E_*. Multiplication by h_0 and h_0^2 leads to the differentials $d_3(h_0^2 \cdot_6 h_2) = h_0 \cdot_0 h_0 d_0$ and $d_3(h_0^3 \cdot_6 h_2) = h_0^2 \cdot_0 h_0 d_0 =_{-1} Pc_0$, respectively. There is no room for further differentials landing in degrees $14 \le t-s \le 15$.

v. Next, we consider differentials landing in degree $t-s=16$. For bidegree reasons the two classes $h_1 \cdot_6 h_2$ and $_1 h_3^2$ in $E_2^{2,18}$ survive to E_∞, and since $\pi_{16}(\mathbb{CP}_{-1}^\infty) \cong \mathbb{Z} \oplus (\mathbb{Z}/2)^2$, the remaining h_0-torsion classes are hit by differentials. Thus $d_2^{1,18}(_5 h_3) = h_0 \cdot_1 h_3^2$ and $d_2^2(h_0 \cdot_5 h_3) = h_0^2 \cdot_1 h_3^2 =_{-1} f_0$.

To determine the last differential landing in degree 16, we compare once again with the Adams spectral sequence $"E_*$ for $\pi_*\left(\mathbb{CP}^\infty\right)$. Comparing the $"E_2$-term and the $"E_\infty$-term given in Table 7.2 of [34] we deduce that there are differentials $"d_2\left(_1h_4\right) =_1 h_0h_3^2$, $"d_3\left(h_0 \cdot_1 h_4\right) =_1 h_0d_0$ and $"d_3\left(h_0^2 \cdot_1 h_4\right) =_1 h_0^2d_0$. In particular, the cited table asserts that $"d_2(_4h_1c_0)=0$ does not interfere with the second $"d_3$-differential. Also $"d_2\left(h_2 \cdot_7 1\right) = h_2 \cdot "d_2\left(_7 1\right) = 0$, and $"d_3\left(h_2 \cdot_7 1\right) = 0$ follows from

$$\pi_{16}^S\left(\mathbb{CP}^\infty\right) \cong \mathbb{Z} \oplus \left(\mathbb{Z}/2\right)^3.$$

The map $j_* : E_2 \to "E_2$ is an isomorphism in degree $t-s=17$ and Adams filtration $s \leq 4$, while in degree $t-s=16$ the kernel consists of the class $h_0^2 \cdot_1 h_3^2 =_{-1} f_0$ only. Comparing $d_2^{1,18}$ to $"d_2^{1,18}$ it follows that $_5h_3 \in E_2^{1,18}$ maps to $_1h_4 \bmod h_2 \cdot_7 1 \in "E_2^{1,18}$. Thus $d_3(h_0^2 \cdot_5 h_3)$ maps under j_* to $_1h_0^2d_0$, and we have proven that $d_3^{3,20}\left(h_0^2 \cdot_5 h_3\right) =_1 h_0^2d_0$ in E_*.

vi. In degree $t-s=17$, the abutment has order 2^8 and the E_2-term has 16 classes. Hence there are five differentials landing in degree 17, in addition to the three differentials we have just found leaving that degree. The 2-torsion in the abutment in degree 18 has order 2^5, and the E_2-term has seven h_0-torsion classes. Hence at most two differentials leave the h_0-torsion in degree 18, and at least three differentials leave the h_0-periodic part of the E_2-term. For bidegree reasons this extreme case is precisely what occurs, so $d_3^{6,24}\left(_9h_0^2\right) \neq 0$, $d_2^{7,25}\left(_9h_0^7\right) =_1 P^2h_2$ and $d_2^{8,26}\left(_9h_0^2\right) =_1 h_0P^2h_2$, and there are no non-zero differentials landing in degree $t-s=18$.

To precisely pin down the differential $d_2^{6,24}$ we use the same argument as for $d_2^{2,12}$. The map $j_* : E_2 \to "E_2$ is an isomorphism in bidegree $(6,24)$ and surjective in bidegree $(8,25)$, so the relation $h_1 \cdot_8 h_0^7 = h_0^7 \cdot_1 h_4$ in $"E_2^{8,25}$ and the differential $"d_2\left(_9h_0^6\right) = h_0^7 \cdot_1 h_0$ implies that $d_2(_9h_0{}^6)$ is either $h_1 \cdot_8 h_0^7$ or $h_0^7 \cdot_5 h_3$ in $E_2^{8,25}$. Multiplying with h_0 and comparing with $d_2^{7,25}$ eliminates the first possibility, so in fact $d_2^{6,24}\left(_9h_0^6\right) = h_0^7 \cdot_5 h_3$.

Considering h_0- and h_2-multiplications in the E_2-term, either $d_2=0$ on all h_0-torsion classes in degree $t-s=18$, or $d_2^{4,22}(x) = h_0^2 \cdot_4 h_1 c_0$ on the classes $x \in E_2^{4,22}$ not divisible by h_0, and $d_2^{5,23}(h_0 \cdot x) = h_0^3 \cdot_4 h_1 c_0$. In the former case, the d_3-differential $d_3^{2,17}$ would propagate by h_2- and h_0-multiplications to three non-zero d_3-differentials from the h_0-torsion in degree $t-s=18$, which is incompatible with the abutment. Thus the two d_2-differentials given above are correct, and this accounts for all the differentials from degree $t-s=18$.

vii. The proofs in degrees $19 \leq t-s \leq 21$ are left as exercises for the reader who needs these results.

THE FIBER OF THE CYCLOTOMIC TRACE MAP

When localized at $p=2$, the homotopy type of the spectrum $K(\mathbb{Z})$ is known. This involves the Bloch–Lichtenbaum spectral sequence relating motivic cohomology to algebraic K-theory, Voevodsky's proof of the Milnor conjecture, which relates motivic cohomology to étale cohomology, and knowledge of the étale cohomology of the rational

2-integers $\mathbb{Z}\left[\dfrac{1}{2}\right]$.

Similarly, the p-adic homotopy type of the spectrum $TC(\mathbb{Z},p)$ is known for each prime p. They were determined by Bökstedt and Madsen in [9] and [10] for p odd, and by the author in [42], [43], [44] and [45] for p=2.

When p=2 the homomorphisms induced by $\mathrm{trc}_{\mathbb{Z}} : K(\mathbb{Z}) \to TC(\mathbb{Z},2)$ on homotopy groups are known after 2-adic completion. In this chapter we use this to describe the homotopy fiber of the cyclotomic trace map as a spectrum.

Let all spectra be implicitly completed at 2, throughout this chapter.

Some Two-Adic K-Theory Spectra

We say that a (-1)-connected spectrum is connective, and a 0-connected spectrum is connected. Let KO and KU denote the real and complex topological K-theory spectra, let ko and ku denote their connective covers, and let bo and bu denote their connected covers, respectively. Write bso and bspin for the 1- and 3-connected covers of KO, and bsu for the 3-connected cover of KU, as usual.

Complex Bott periodicity provides a homotopy equivalence $\beta : \Sigma^2 KU \to KU$. There is a complexification map $c : KO \to KU$ and a realification map $r : KU \to KO$. Smashing with the Hopf map $\eta : \Sigma^\infty S^{-1} \to \Sigma^\infty S^0$ yields a map also denoted $\eta : \Sigma KO \to KO$. We use the same notation for the various k-connected covers of these maps. There is a cofiber sequence of spectra

$$\Sigma ko \xrightarrow{\eta} ko \xrightarrow{c} ku \xrightarrow{\Sigma^2 r \circ \beta^{-1}} \Sigma^2 ko. \tag{3.2}$$

This follows from R. Wood's theorem $KO \wedge \mathbb{CP}^2 \simeq KU$. See also [30, V.5.15]. Here we write $\Sigma^2 r \circ \beta^{-1}$ for a map $\partial : ku \to \Sigma^2 Ko$ that satisfies $\partial \circ \beta = \Sigma^2 r$. This determines the map up to homotopy, even though $\beta : \Sigma^2 ku \to ku$ is not exactly invertible.

Theorem 3.3 Quillen

There is a cofiber sequence of spectra

$$K(\mathbb{F}_3) \xrightarrow{i_3} ku \xrightarrow{\psi^3 - 1} bu \xrightarrow{\partial_3} \Sigma K(\mathbb{F}_3).$$

This is the spectrum level statement of Quillen's computation in [38].

The computation in [46] by Weibel and the author of the 2-primary algebraic K-groups of rings of 2-integers in number fields relies on Suslin's motivic cohomology for fields [49], Voevodsky's proof of the Milnor conjecture [51] and the Bloch–Lichtenbaum spectral sequence [4].

In the case of the 2-integers $\mathbb{Z}\left[\dfrac{1}{2}\right]$ in $\mathbb{Q}$ the result implies that there is

a 2-adic homotopy equivalence $K\left(\mathbb{Z}\left[\dfrac{1}{2}\right]\right) \simeq JK\left(\mathbb{Z}\left[\dfrac{1}{2}\right]\right)$, where the latter spectrum was defined by Bökstedt in [5]. This leads to the following statement:

Theorem 3.4 Rognes–Weibel

There is a cofiber sequence of spectra

$$\Sigma ko \longrightarrow K(\mathbb{Z}[\tfrac{1}{2}]) \overset{\pi_3}{\longrightarrow} K(\mathbb{F}_3) \overset{\partial}{\longrightarrow} \Sigma^2 ko,$$

Where π_3 is induced by the ring surjection $\pi_3 : \mathbb{Z}\left[\dfrac{1}{2}\right] \to \mathbb{F}_3$. The connecting map ∂ is homotopic to the composite

$$K(\mathbb{F}_3) \overset{i_3}{\longrightarrow} ku \overset{\Sigma^2 r \circ \beta^{-1}}{\longrightarrow} \Sigma^2 ko.$$

Proof

Bökstedt's $JK\left(\mathbb{Z}\left[\dfrac{1}{2}\right]\right)$ can be defined as the homotopy fiber of the composite

$$ko \overset{c}{\longrightarrow} ku \overset{\psi^3 - 1}{\longrightarrow} bu.$$

By Bökstedt [5, Theorem 2] there is a map $\phi : K\left(\mathbb{Z}\left[\frac{1}{2}\right]\right) \to JK\left(\mathbb{Z}\left[\frac{1}{2}\right]\right)$ inducing a split surjection on homotopy. By Rognes and Weibel [46] and Weibel [60] these spectra have isomorphic homotopy groups, hence Φ is a homotopy equivalence. There is a square of horizontal and vertical cofiber sequences:

The left hand vertical yields the asserted cofiber sequence. Bökstedt›s cited construction of the map Φ identifies the composite $K\left(\mathbb{Z}\left[\frac{1}{2}\right]\right) \to JK\left(\mathbb{Z}\left[\frac{1}{2}\right]\right) \to K(\mathbb{F}_3)$ with that induced by the ring homomorphism π_3.

The Reduction Map

Let us recall the Galois reduction map from [18, Section 13; 45, Section 3]. Let $\phi^3 \in \text{Gal}\left(\overline{\mathbb{Q}}_2 / \mathbb{Q}_2\right)$ be a Galois automorphism of the algebraic closure $\overline{\mathbb{Q}}_2$ of the field $\mathbb{Q}_2$ of 2-adic numbers, such that $\varphi^3(\zeta) = \zeta^3$ when ζ is a 2-power root of unity, i.e., in $\mu_{2^\infty} \subset \overline{\mathbb{Q}}_2^\times$. We may further assume that $\phi^3\left(+\sqrt{3}\right) = +\sqrt{3}$. Then φ^3 induces a self-map of $K\left(\overline{\mathbb{Q}}_2\right)$ which is compatible up to homotopy with $\psi^3 : ku \to ku$ under Suslin's (implicitly 2-adic) homotopy equivalence $K\left(\overline{\mathbb{Q}}_2\right) \simeq ku$ from [48]. Hence, the inclusion $K(\mathbb{Q}_2) \to K\left(\overline{\mathbb{Q}}_2\right)^{h\phi^3}$ to the homotopy equalizer of φ^3 and the identity on $K\left(\overline{\mathbb{Q}}_2\right)$ yields a spectrum map $K(\mathbb{Q}_2) \to (ku)^{h\psi^3}$. The connective cover of the target is identified with $K(\mathbb{F}_3)$ by Quillen's theorem, which defines the Galois reduction map

$$\text{red} : K(\mathbb{Q}_2) \to K(\mathbb{F}_3).$$

Fix a choice of such a map.

Theorem 3.6 Rognes

There are cofiber sequences of spectra

$$K^{\text{red}}(\mathbb{Q}_2) \longrightarrow K(\mathbb{Q}_2) \xrightarrow{\text{red}} K(\mathbb{F}_3) \xrightarrow{\partial_2} \Sigma K^{\text{red}}(\mathbb{Q}_2)$$

And

$$\Sigma K(\mathbb{F}_3) \xrightarrow{f_{\mathbb{C}}} K^{\text{red}}(\mathbb{Q}_2) \longrightarrow \Sigma ku \xrightarrow{\partial_1} \Sigma^2 K(\mathbb{F}_3).$$

The former connecting map ∂_2 is determined by its composite with $\Sigma K^{\text{red}}(\mathbb{Q}_2) \to \Sigma^2 ku$, which up to a two-adic unit is homotopic to the composite

$$K(\mathbb{F}_3) \xrightarrow{i_3} ku \xrightarrow{1-\psi^{-1}} bu \xrightarrow[\simeq]{\beta^{-1}} \Sigma^2 ku.$$

The latter connecting map ∂_1 is homotopic to the composite

$$\Sigma ku \xrightarrow{\Sigma(1-\psi^{-1})} \Sigma bu \xrightarrow{\Sigma \partial_3} \Sigma^2 K(\mathbb{F}_3).$$

Both connecting maps induce the zero map on homotopy, and the extensions

$$K_*^{\text{red}}(\mathbb{Q}_2) \longrightarrow K_*(\mathbb{Q}_2) \xrightarrow{\text{red}_*} K_*(\mathbb{F}_3)$$

and

$$\pi_*(\Sigma K(\mathbb{F}_3)) \xrightarrow{f_{\mathbb{C}*}} K_*^{\text{red}}(\mathbb{Q}_2) \longrightarrow \pi_*(\Sigma ku)$$

are both split.

This is the conclusion of [45, 8.1]. To go on, we consider the ring homomorphisms $j: \mathbb{Z} \to \mathbb{Z}_2$ and $j': \mathbb{Z}\left[\frac{1}{2}\right] \to \mathbb{Q}_2$.

Theorem 3.7 Quillen

There is a map of horizontal cofiber sequences of spectra

Inducing a homotopy equivalence $\mathrm{hofib}(j) \to \mathrm{hofib}(j')$.

This is the spectrum level statement of the localization sequences in K-theory from [39].

Theorem 3.8 Hesselholt–Madsen

In the commutative square of spectra

$$
\begin{array}{ccc}
K(\mathbb{Z}) & \xrightarrow{\ j\ } & K(\mathbb{Z}_2) \\
\downarrow{\scriptstyle \mathrm{trc}_{\mathbb{Z}}} & & \downarrow{\scriptstyle \mathrm{trc}_{\mathbb{Z}_2}} \\
TC(\mathbb{Z}) & \xrightarrow[\simeq]{\ j\ } & TC(\mathbb{Z}_2)
\end{array}
$$

The right hand map induces a homotopy equivalence on connective covers. The lower map is a homotopy equivalence, and there is a cofiber sequence of spectra

$$
\mathrm{hofib}(j) \to \mathrm{hofib}(\mathrm{trc}_{\mathbb{Z}}) \to \Sigma^{-2} H\mathbb{Z}.
$$

This is Theorem D of [22], which uses McCarthy's theorem [31].

Theorem 3.9 Rognes

The natural map $j' : K\left(\mathbb{Z}\left[\frac{1}{2}\right]\right) \to K(\mathbb{Q}_2)$ induces an isomorphism of 2-adic homotopy groups modulo torsion, in each positive dimension $* \equiv 1 \bmod 4$.

This is the content of [45, 7.7]. By a homotopy group modulo torsion we here mean the quotient of the homotopy group by its torsion subgroup. Hence, the assertion is stronger than just saying that j' induces a homomorphism whose kernel and cokernel are torsion groups.

Proposition 3.10

There is a map of horizontal cofiber sequences of spectra

$$
\begin{array}{ccccc}
\Sigma ko & \longrightarrow & K(\mathbb{Z}[\tfrac{1}{2}]) & \xrightarrow{\pi_3} & K(\mathbb{F}_3) \\
\downarrow {\scriptstyle j^{\mathrm{red}}} & & \downarrow {\scriptstyle j'} & & \simeq \downarrow {\scriptstyle \bar{j}} \\
K^{\mathrm{red}}(\mathbb{Q}_2) & \longrightarrow & K(\mathbb{Q}_2) & \xrightarrow{\mathrm{red}} & K(\mathbb{F}_3)
\end{array}
$$

Such that the right hand map $\bar{j}$ is a homotopy equivalence. Hence there is a homotopy equivalence $\mathrm{hofib}(j^{\mathrm{red}}) \xrightarrow{\simeq} \mathrm{hofib}(j')$.

Proof

Suppose we have shown that the composite

$$
\Sigma ko \to K(\mathbb{Z}[\tfrac{1}{2}]) \xrightarrow{j'} K(\mathbb{Q}_2) \xrightarrow{\mathrm{red}} K(\mathbb{F}_3)
$$

Is null homotopic. Then a choice of null homotopy defines an extension $\bar{J} : K(\mathbb{F}_3) \to K(\mathbb{F}_3)$ of red∘j' over π_3, as well as a lifting $j^{\mathrm{red}} : \Sigma ko \to K^{\mathrm{red}}(\mathbb{Q}_2)$. By the calculations of [45, 4.2 and 4.4] the composite red∘j' is surjective on homotopy in dimensions $0 \le * \le 7$, hence in all dimensions by $v_1{}^4$-periodicity. Thus $\bar{j}$ induces surjections on homotopy in all dimensions, and must be a homotopy equivalence. This will then complete the proof of the proposition.

To show that the composite map $\Sigma ko \to K(\mathbb{F}_3)$ is null homotopic, it suffices to show that precomposition with the connecting map $\partial : K(\mathbb{F}_3) \to \Sigma^2$ in Theorem 3.4 induces an injection between the groups of homotopy classes of maps to $\Sigma K(\mathbb{F}_3)$:

$$[\Sigma^2 ko, \Sigma K(\mathbb{F}_3)] \xrightarrow{\partial^\#} [K(\mathbb{F}_3), \Sigma K(\mathbb{F}_3)].$$

Here ∂ is the composite of $i_3 : K(\mathbb{F}_3) \to ku$ with the map denoted $\Sigma^2 ro \beta^{-1} : ku \to \Sigma^2 ko$. Thus it suffices to show that both homomorphisms $i_3^\#$ and $(\Sigma^2 ro\beta^{-1})^\#$ are injective.

There is an exact sequence

$$[bu, \Sigma K(\mathbb{F}_3)] \xrightarrow{(\psi^3 - 1)^\#} [ku, \Sigma K(\mathbb{F}_3)] \xrightarrow{i_3^\#} [K(\mathbb{F}_3), \Sigma K(\mathbb{F}_3)].$$

Any map $bu \to \Sigma K(\mathbb{F}_3)$ has the form $\partial_3 \circ \varphi$, for some operation $\phi : bu \to bu$. Thus its precomposition with $(\psi^3 - 1)$ is null homotopic, because φ and $(\psi^3 - 1)$ commute and $\partial_3 \circ (\psi^3 - 1) \simeq *$. Hence the left hand map is null and $i_3^\#$ is injective.

There is also an exact sequence

$$[\Sigma ko, \Sigma K(\mathbb{F}_3)] \xrightarrow{\eta^\#} [\Sigma^2 ko, \Sigma K(\mathbb{F}_3)] \xrightarrow{(\Sigma^2 ro\beta^{-1})^\#} [ku, \Sigma K(\mathbb{F}_3)].$$

From Theorem 3.3 we see that $\left[\Sigma ko, \Sigma K(\mathbb{F}_3)\right]$ is zero, because postcomposition with $(\psi^3 - 1)$ acts injectively on the homotopy classes of maps $ko \to ku$, see [29]. Thus also $(\Sigma^2 ro\beta^{-1})^\#$ is injective, which completes the proof.

Fix choices of maps j^{red} and $\bar{j}$, as above.

Proposition 3.11

There is a cofiber sequence of spectra

$$K(\mathbb{F}_3) \to \text{hofib}(j^{red}) \to \Sigma^2 ko \xrightarrow{\partial} \Sigma K(\mathbb{F}_3).$$

The connecting map ∂ is homotopic to the composite

$$\Sigma^2 ko \xrightarrow{\Sigma^2 c} \Sigma^2 ku \xrightarrow[\sim]{\beta} bu \xrightarrow{\partial_3} \Sigma K(\mathbb{F}_3).$$

Proof

The map $\Sigma ko \to K\left(\mathbb{Z}\left[\frac{1}{2}\right]\right)$ induces an isomorphism on 2-adic homotopy modulo torsion in dimensions $* \equiv 1 \bmod 8$, and multiplication by 2 times a 2-adic unit in dimensions $* \equiv 5 \bmod 8$. By Theorem 3.9 the same holds for the composite map from Σko to $K(\mathbb{Q}_2)$, and by Theorem 3.6 the same holds for the lift $J^{red} : \Sigma ko \to K^{red}(\mathbb{Q}_2)$, as well as the composite map $\Sigma ko \to K^{red}(\mathbb{Q}_2) \to \Sigma ku$. Any such map factors as a self-map φ of Σko followed by the suspended complexification map $\Sigma c \to Ko \to \Sigma ku$. Since the suspended complexification map induces the identity in dimensions $* \equiv 1 \bmod 8$, and multiplication by 2 in dimensions $* \equiv 5 \bmod 8$, it follows that φ is a 2-adic homotopy equivalence. We obtain the following diagram of horizontal and vertical cofiber sequences:

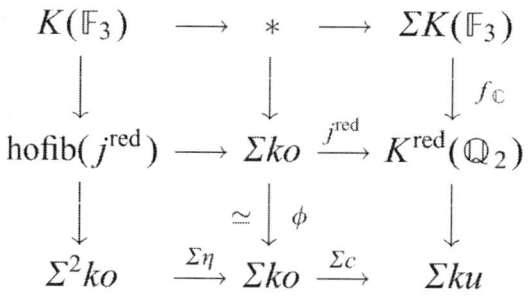

The connecting map $\partial : \Sigma^2 ko \to \Sigma K(\mathbb{F}_3)$ is detected by its precomposition with $\Sigma^2 ro \, \beta^{-1} : ku \to \Sigma^2 ko$, because $[\Sigma ko, \Sigma K(\mathbb{F}_3)] = 0$. By the diagram above, the composite $\partial \circ \Sigma^2 ro \beta^{-1}$ is the desuspended connecting map $\Sigma^{-1}\partial_1 = \partial_3 \circ (1 - \psi^{-1})$ from 3.6. Thus $\partial = \partial_3 \circ \beta \circ \Sigma^2 c$ in the stable category, by the calculation

$$\partial_3 \circ \beta \circ \Sigma^2 c \circ \Sigma^2 r \circ \beta^{-1} = \partial_3 \circ (1 - \psi^{-1})$$

Two-Primary Algebraic K-Theory of Pointed Spaces

Which uses $\text{cor} = 1 + \psi^{-1}$, and $\psi k \circ \beta = \beta \circ \Sigma^2(k\psi k)$.

Proposition 3.12

There is a cofiber sequence of spectra

$$\Sigma^3 ko \longrightarrow \text{hofib}(j^{\text{red}}) \longrightarrow ku \overset{\partial}{\longrightarrow} \Sigma^4 ko.$$

The connecting map ∂ is homotopic to the composite

$$ku \overset{\psi^3 - 1}{\longrightarrow} bu \underset{\simeq}{\overset{\beta^{-1}}{\longrightarrow}} \Sigma^2 ku \overset{\Sigma^2(\Sigma^2 r \circ \beta^{-1})}{\longrightarrow} \Sigma^4 ko$$

with the same notation as in (3.2). It is characterized by the following homotopy commutative diagram:

$$
\begin{array}{ccc}
ku & \overset{\partial}{\longrightarrow} & \Sigma^4 ko \\
\downarrow{\scriptstyle cov} & & \downarrow{\scriptstyle cov} \\
KU & & \Sigma^4 KO
\end{array}
$$

$$KU \overset{\psi^3 - 1}{\longrightarrow} KU \underset{\simeq}{\overset{\beta^2}{\longleftarrow}} \Sigma^4 KU \overset{\Sigma^4 r}{\longrightarrow} \Sigma^4 KO.$$

The maps labeled cov are k-connective covering maps, for suitable k.

Proof

We use the factorization of the connecting map in Proposition 3.11 to form the following diagram of horizontal and vertical cofiber sequences:

Table 4a: The Adams E_2-term for hofib(trc)

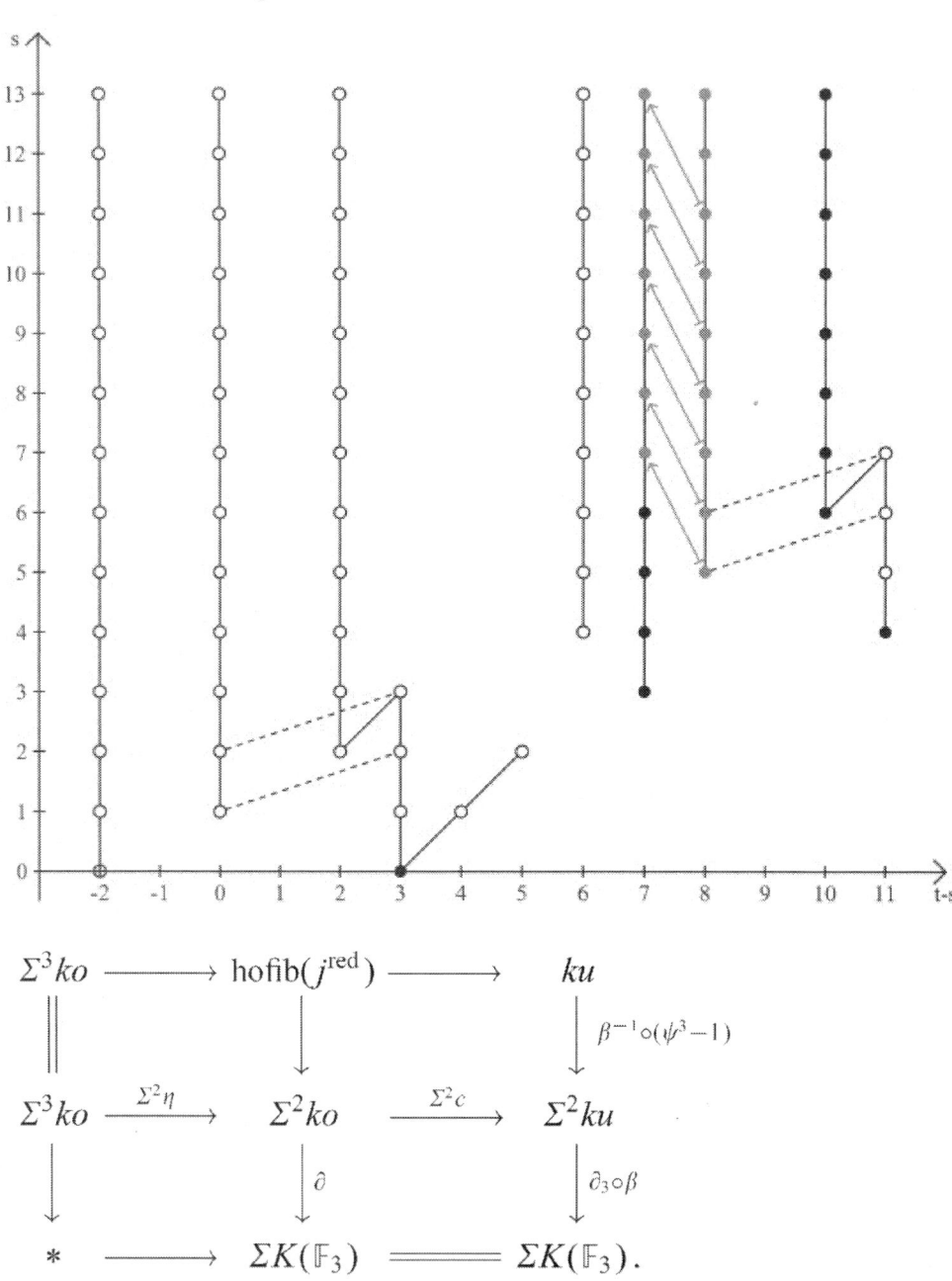

Two-Primary Algebraic K-Theory of Pointed Spaces

Table 4b: The Adams E_2-term for hofib(trc)

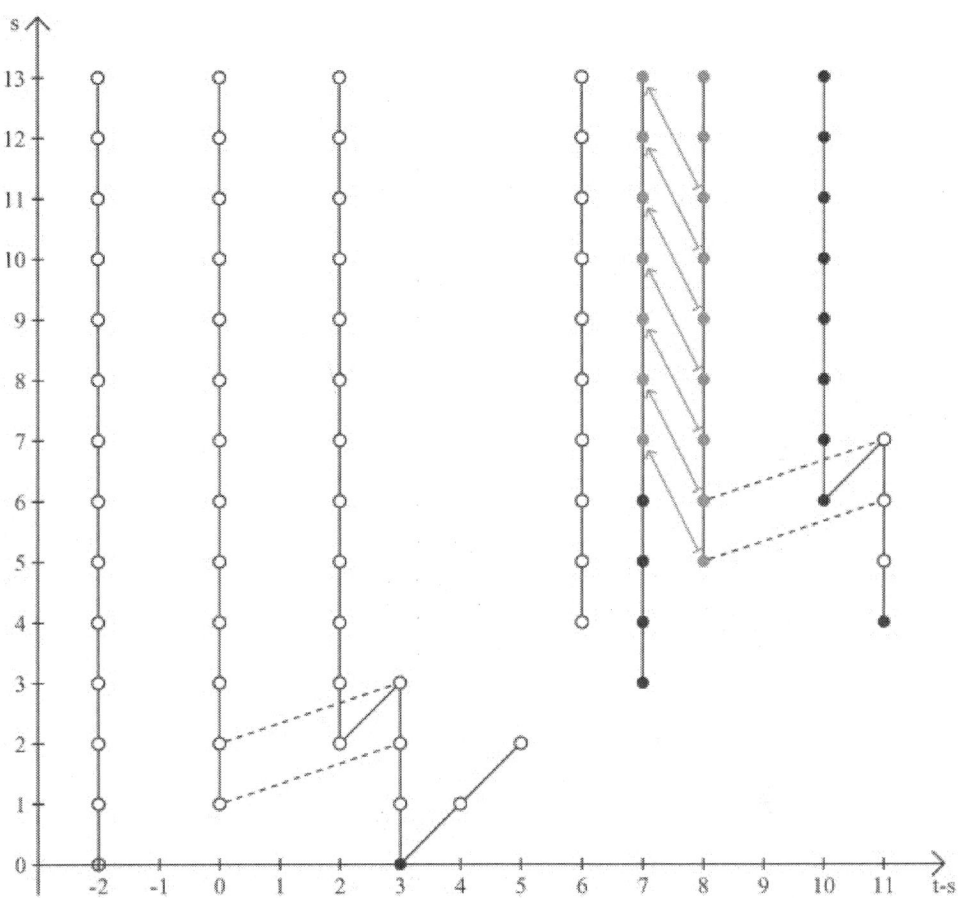

The right hand column is a variant of the sequence in Theorem 3.3. It follows that the connecting map $ku \to \Sigma^4 ko$ for the top row is the composite of

$$ku \xrightarrow{\psi^3 - 1} bu \xrightarrow[\simeq]{\beta^{-1}} \Sigma^2 ku$$

and the connecting map for the middle row, i.e., the double suspension of the connecting map $\Sigma^2 ro\, \beta^{-1} = \partial : ku \to \Sigma^2 ko$ of (3.2).

The covering maps induce an injection $\mathrm{cov}_* : \left[ku, \Sigma^4 ko \right] \to \left[ku, \Sigma^4 KO \right]$ and a bijection $\mathrm{cov}^\# : \left[KU, \Sigma^4 KO \right] \cong \left[ku, \Sigma^4 KO \right]$, and the connecting map ∂ corresponds to $\Sigma^4 r \circ \beta^{-2} \circ (\psi^3 - 1)$ in $[KU, \Sigma^4 KO]$. Hence ∂ is characterized by the given diagram.

The following theorem is the main result of this chapter.

Theorem 3.13

There is a cofiber sequence of spectra

$$\Sigma^3 ko \to \mathrm{hofib}(\mathrm{trc}) \to \Sigma^{-2} ku \xrightarrow{\delta} \Sigma^4 ko.$$

The connecting map δ is characterized by the following homotopy commutative diagram:

$$
\begin{array}{ccc}
\Sigma^{-2} ku & \xrightarrow{\quad\quad\quad\delta\quad\quad\quad} & \Sigma^4 ko \\
\downarrow{\scriptstyle cov} & & \downarrow{\scriptstyle cov} \\
\Sigma^{-2} KU \xleftarrow[\simeq]{\beta} KU \xrightarrow{\psi^3-1} KU \xleftarrow[\simeq]{\beta^2} \Sigma^4 KU & \xrightarrow{\Sigma^4 r} & \Sigma^4 KO.
\end{array}
$$

The maps labeled cov are suitable covering maps.

Proof

Consider the following diagram of horizontal and vertical cofiber sequences of spectra, obtained by combining Theorem 3.7 and Theorem 3.8 and Proposition 3.10 and Proposition 3.12

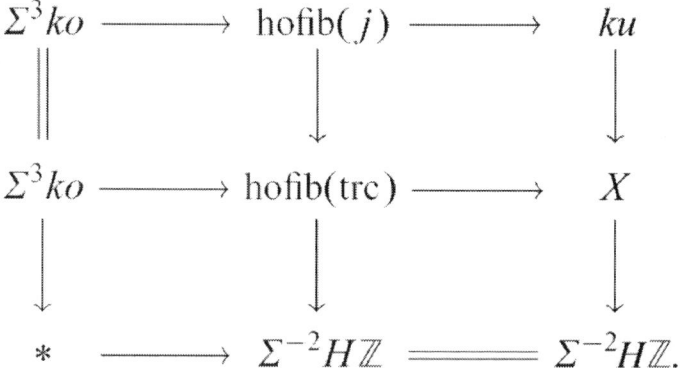

Here X is the cofiber of the composite map $\Sigma^3 ko \to$ hofib (j) $\to$ hofib(trc). It is classified as an extension of $\Sigma^{-2} H\mathbb{Z}$ by ku by an element in $\left[\Sigma^{-2} H\mathbb{Z}, \Sigma ku\right] \cong \mathbb{Z}_2$, whose mod 2 reduction is detected by the composite $k: \Sigma^{-2} H\mathbb{Z} \to \Sigma ku \to \Sigma H\mathbb{Z}$ in $\left[\Sigma^{-2} H\mathbb{Z}, \Sigma ku\right] \cong \mathbb{Z}_2 / 2$. (Here $\Sigma ku \to \Sigma H\mathbb{Z}$ is the map inducing an isomorphism on π_1.) This composite k is the k-invariant of X relating the homotopy groups in dimensions -2 and 0.

Since $\Sigma^3 ko$ is 2-connected, this lowest k-invariant is the same for X as for hofib(trc). By combining Theorem 1.11 with Corollary 1.21 we obtain a cofiber sequence

$$\mathbb{C}P_{-1}^\infty \xrightarrow{i} \text{hofib(trc)} \xrightarrow{j} Wh^{\text{Diff}}(*) \qquad 3.14$$

Whose connecting map is identified with $\widetilde{\text{trc}}$. Since $Wh^{\text{Diff}}(*)$ is connected, it follows that the lowest k-invariants for hofib(trc) and $\mathbb{C}P_{-1}^\infty$ are equal. By Lemma 2.14 the latter is non-zero. Hence k is the essential map.

It follows that X is classified by a map $u \cdot \partial$ where $\partial: \Sigma^{-2} H\mathbb{Z} \to \Sigma ku$ classifies $\Sigma^{-2}ku$ and u is a 2-adic unit. We get a homotopy equivalence of cofiber sequences:

Hence $X \simeq \Sigma^{-2} ku$, as claimed.

To characterize δ, we compare with the connecting map ∂ of Proposition 3.12. Precomposition with $\beta : ku \to \Sigma^{-2}$, or its K-localization, induces the vertical maps in the following commutative diagram:

$$
\begin{array}{ccccc}
[\Sigma^{-2} ku, \Sigma^4 ko] & \xrightarrow{cov_\#} & [\Sigma^{-2} ku, \Sigma^4 KO] & \xleftarrow{cov^\#}_{\cong} & [\Sigma^{-2} KU, \Sigma^4 KO] \\
\downarrow \beta^\# & & \cong \downarrow \beta^\# & & \cong \downarrow \beta^\# \\
[ku, \Sigma^4 ko] & \xrightarrow{cov_\#} & [ku, \Sigma^4 KO] & \xleftarrow{cov^\#}_{\cong} & [KU, \Sigma^4 KO].
\end{array}
$$

Here the maps labeled $cov_\#$ are injective, and the maps labeled $cov^\#$ are bijective. The class δ in $[\Sigma^{-2} ku, \Sigma^4 ko]$ maps to ∂ under $\beta^\#$, which in turn maps to $\Sigma^4 r \circ \beta^{-2} \circ (\psi^3 - 1)$ in $[KU, \Sigma^4 KO]$ by Proposition 3.12. The right hand $\beta^\#$ is bijective, so this characterizes the image of δ in $[\Sigma^{-2} KU, \Sigma^4 KO]$ as $\Sigma^4 r \circ \beta^{-2} \circ (\psi^3 - 1) \circ \beta^{-1}$. This characterizes δ up to homotopy, by the injectivity claims above.

Remark 3.15

By Bökstedt et al. [7] or Corollary 1.8 this theorem also determines the homotopy fiber of the cyclotomic trace map $trc_X : A(X) \to TC(X)$ completed at 2 for any 1-connected space X, since the natural map

$$
hofib(trc_X) \xrightarrow{\;\simeq\;} hofib(trc)
$$

is a homotopy equivalence.

Let $v_2(k)$ be the 2-adic valuation of k.

Corollary 3.16

In positive dimensions (n>0) the homotopy groups of hofib(trc) are

$$\pi_n(\mathrm{hofib}(\mathrm{trc})) \cong \begin{cases} 0 \text{ for } n \equiv 0,\ 1 \bmod 8, \\[1ex] \mathbb{Z} \text{ for } n \equiv 2, 6 \bmod 8, \\[1ex] \mathbb{Z}/16 \text{ for } n \equiv 3 \bmod 8, \\[1ex] \mathbb{Z}/2 \text{ for } n \equiv 4, 5 \bmod 8, \\[1ex] \mathbb{Z}/2^{v_2(k)+4} \text{ for } n = 8k - 1. \end{cases}$$

Also $\pi_n(\mathrm{hobic}(\mathrm{trc})) \cong \mathbb{Z}$ for $n=-2$ and $n=0$. The remaining homotopy groups are zero.

Proof

This is a routine calculation, given the action of ψ^3-1 and $\Sigma^4 r$ on homotopy.

The spectrum cohomology of hofib(trc) is given in 4.4 below. The Adams E_2-term

equation(3.17)

$$E_2^{s,t} = \mathrm{Ext}_A^{s,t}(H^*(\mathrm{hofib}(\mathrm{trc})), \mathbb{F}_2) \Rightarrow \pi_{t-s}(\mathrm{hofib}(\mathrm{trc}))_2^\wedge$$

is then easily deduced from the E_2-terms in the Adams spectral sequences for $\pi_*(\mathrm{ko})_2^\wedge$ and $\pi_*(\mathrm{ku})_2^\wedge$. Furthermore, only one pattern of differentials is compatible with Corollary 3.16: There is an infinite h_0-tower of non-zero d_r-differentials from column $t-s=8k$ for all $k \geq 1$, with $r=v_2(k)+2$, and no other differentials. The spectral sequence is displayed in Table 4a and Table 4b.

COHOMOLOGY OF THE SMOOTH WHITEHEAD SPECTRUM

We now determine the mod2 spectrum cohomology of the smooth Whitehead spectrum of a point, as a module over the Steenrod algebra.

Consider the following diagram:

$$
\begin{array}{ccc}
\mathbb{C}P^{\infty}_{-1} & =\!=\!=\!=\!=\!= & \mathbb{C}P^{\infty}_{-1} \\
\downarrow {\scriptstyle i} & & \downarrow {\scriptstyle \varepsilon} \\
\Sigma^{3}ko \longrightarrow \mathrm{hofib(trc)} \longrightarrow \Sigma^{-2}ku \xrightarrow{\ \delta\ } \Sigma^{4}ko \\
\| \qquad\qquad\quad \downarrow {\scriptstyle j} \qquad\qquad\quad \downarrow \qquad\qquad\quad \| \\
\Sigma^{3}ko \longrightarrow \mathrm{Wh}^{\mathrm{Diff}}(*) \longrightarrow \Sigma\,\mathrm{hofib}(\varepsilon) \longrightarrow \Sigma^{4}ko\,.
\end{array} \qquad (4.1)
$$

The middle row is the cofiber sequence from Theorem 3.13, and the left column is (3.14). We let ε be the composite map $\mathbb{C}P^{\infty}_{-1} \to \mathrm{hobit(trc)} \to \Sigma^{-2}ku$. Then the right column and bottom row are also cofiber sequences.

Proposition 4.2

The mapεinduces the unique surjection of$A$-modules

$$\Sigma^{-2}A/A(Sq^{1},Sq^{3}) \cong H^{*}(\Sigma^{-2}ku) \xrightarrow{\varepsilon^{*}} H^{*}(\mathbb{C}P^{\infty}_{-1}) \cong \Sigma^{-2}A/C.$$

Hence

$$H^{*}(\Sigma\,\mathrm{hofib}(\varepsilon)) \cong \Sigma^{-2}C/A(Sq^{1},Sq^{3})$$

as an$A$-module.

Two-Primary Algebraic K-Theory of Pointed Spaces

Table 5: $H^*(\mathrm{Wh}^{\mathrm{Diff}}(*))$ in dimensions ≤ 14

dim	x	$Sq^1(x)$	$Sq^2(x)$	$Sq^4(x)$	$Sq^8(x)$
3	ι_3	$Sq^4Sq^2\iota_{-2}$	$Sq^7\iota_{-2}$	$Sq^4\iota_3$	$Sq^8\iota_3$
4	$Sq^4Sq^2\iota_{-2}$	0	$Sq^6Sq^2\iota_{-2}$	0	$Sq^8Sq^4Sq^2\iota_{-2}$
5	$Sq^7\iota_{-2}$	0	$Sq^9\iota_{-2}$ $\equiv Sq^7Sq^2\iota_{-2}$	$Sq^{11}\iota_{-2}$	$Sq^{13}Sq^2\iota_{-2}$ $+Sq^{11}Sq^4\iota_{-2}$
6	$Sq^6Sq^2\iota_{-2}$	$Sq^9\iota_{-2}$ $\equiv Sq^7Sq^2\iota_{-2}$	0	$Sq^{10}Sq^2\iota_{-2}$	$Sq^{10}Sq^4Sq^2\iota_{-2}$
7	$Sq^9\iota_{-2}$ $\equiv Sq^7Sq^2\iota_{-2}$ $Sq^4\iota_3$	0 0	0 $Sq^6\iota_3$	$Sq^{11}Sq^2\iota_{-2}$ $Sq^{13}\iota_{-2}$	
8	$Sq^8Sq^2\iota_{-2}$	0	$Sq^{10}Sq^2\iota_{-2}$	$Sq^{12}Sq^2\iota_{-2}$	
9	$Sq^{11}\iota_{-2}$ $Sq^6\iota_3$	0 $Sq^7\iota_3$	$Sq^{13}\iota_{-2}$ 0	$Sq^{13}Sq^2\iota_{-2}$ $Sq^{13}Sq^2\iota_{-2}$ $+Sq^{11}Sq^4\iota_{-2}$ $+Sq^{10}\iota_3$	
10	$Sq^{10}Sq^2\iota_{-2}$ $Sq^8Sq^4\iota_{-2}$ $Sq^7\iota_3$	$Sq^{11}Sq^2\iota_{-2}$ 0 0	0 $Sq^{10}Sq^4\iota_{-2}$ 0	0 $Sq^{12}Sq^4\iota_{-2}$ $Sq^{11}\iota_3$	
11	$Sq^{13}\iota_{-2}$ $Sq^{11}Sq^2\iota_{-2}$ $Sq^8\iota_3$	0 0 $Sq^8Sq^4Sq^2\iota_{-2}$	0 $Sq^{13}Sq^2\iota_{-2}$ $Sq^{10}\iota_3$		
12	$Sq^{12}Sq^2\iota_{-2}$ $Sq^{10}Sq^4\iota_{-2}$ $Sq^8Sq^4Sq^2\iota_{-2}$	$Sq^{13}Sq^2\iota_{-2}$ $Sq^{11}Sq^4\iota_{-2}$ 0	$Sq^{14}Sq^2\iota_{-2}$ 0 $Sq^{10}Sq^4Sq^2\iota_{-2}$		
13	$Sq^{15}\iota_{-2}$ $Sq^{13}Sq^2\iota_{-2}$ $Sq^{11}Sq^4\iota_{-2}$ $Sq^{10}\iota_3$	0 0 0 $Sq^{11}\iota_3$			
14	$Sq^{14}Sq^2\iota_{-2}$ $Sq^{12}Sq^4\iota_{-2}$ $Sq^{10}Sq^4Sq^2\iota_{-2}$ $Sq^{11}\iota_3$				

Proof

We use that Σ^4ko and $\text{Wh}^{\text{Diff}}(*)$ are connective spectra. Hence $\sum\text{hofib}(\varepsilon)$ is connective, and so ε induces an isomorphism in dimension -2. This determines ε^* since $H^*(\Sigma^{-2}\text{ku})$ is a cyclic A-module, and ε^* is surjective because $H^*(\mathbb{C}P_{-1}^\infty)$ is a cyclic A-module. We identify $H^*(\sum\text{hofib}(\varepsilon))$ with $\ker(\varepsilon^*)$.

Proposition 4.3

The connecting map δ induces the zero homomorphism on cohomology.

Proof

In fact, the group of A-module homomorphisms

$H^*(\Sigma^4\text{ko}) \cong \Sigma^4 A/A(Sq^1, Sq^2) \to \Sigma^{-2}A/A(Sq^1, Sq^3) \cong H^*(\Sigma^{-2}\text{ku})$ is zero. For $A/A(Sq^1, Sq^3)$ is $F_2\{Sq^6; Sq^4Sq^2\}$ in dimension 6, while $Sq^1 \circ Sq^6 \neq 0$ and $Sq^2 \circ Sq^4Sq^2 \neq 0$ in this A-module.

Theorem 4.4

The mod 2 spectrum cohomology of hofib(trc) is the unique non-trivial extension of A-modules

$$\Sigma^{-2}A/A(Sq^1, Sq^3) \to H^*(\text{hofib}(\text{trc})) \to \Sigma^3 A/A(Sq^1, Sq^2).$$

Theorem 4.5

The mod 2 spectrum cohomology of $\text{Wh}^{\text{Diff}}(*)$ is the unique non-trivial extension of A-modules

$$\Sigma^{-2}C/A(Sq^1, Sq^3) \to H^*(\text{Wh}^{\text{Diff}}(*)) \to \Sigma^3 A/A(Sq^1, Sq^2).$$

The mod2 spectrum cohomology of $A(*)$ is given by the splitting of A-modules

$$H^*(A(*)) \cong H^*(\mathrm{Wh}^{\mathrm{Diff}}(*)) \oplus \mathbb{F}_2.$$

Here $\mathbb{F}_2 = H^*(S^0)$ denotes the trivial A-module concentrated in degree zero. We prove these two theorems together.

Proof of Theorem 4.4 and Theorem 4.5

We apply mode 2 spectrum cohomology to (4.1). By Proposition 4.2 the map ε induces a surjection in each dimension, so $\Sigma^{-2} ku \to \mathrm{hofib}(\varepsilon)$ induces an injection in each dimension. By Proposition 4.3 the map δ induces the zero homomorphism in each dimension, and combining these facts we see that $\Sigma\,\mathrm{hofib}(\varepsilon) \to \Sigma^{-2} ko$ also induces the zero homomorphism in cohomology. Thus, the long exact sequences in cohomology associated to the middle and lower horizontal cofiber sequences in (4.1) break up into short exact sequences. These express $H^*(\mathrm{hobit}(\mathrm{trc}))$ and $H^*\mathrm{Wh}^{\mathrm{Diff}}(*)$ as extensions of A-modules, as claimed.

$$
\begin{array}{ccc}
\Sigma^{-2}A/C & =\!=\!=\!= & \Sigma^{-2}A/C \\
\uparrow{\scriptstyle i^*} & & \uparrow{\scriptstyle \varepsilon^*} \\
\Sigma^3 A/A(Sq^1,Sq^2) \;\longleftarrow\; H^*(\mathrm{hofib}(\mathrm{trc})) \;\longleftarrow\; \Sigma^{-2}A/A(Sq^1,Sq^3) \\
\| \qquad\qquad\qquad \uparrow{\scriptstyle j^*} \qquad\qquad\qquad \uparrow \\
\Sigma^3 A/A(Sq^1,Sq^2) \;\longleftarrow\; H^*(\mathrm{Wh}^{\mathrm{Diff}}(*)) \;\longleftarrow\; \Sigma^{-2}C/A(Sq^1,Sq^3)
\end{array}
\qquad (4.6)
$$

It remains to characterize the extensions, which are represented by elements of Ext_A^1. Recall that $H^*(ko) = A/A(Sq^1,Sq^2) = A//A_1$ where $A_1 \subset A$ is the sub-Hopf algebra generated by Sq^1 and Sq^2. Hence, there are change-of-rings isomorphism's

$$\mathrm{Ext}_A^1(\Sigma^3 A//A_1, \Sigma^{-2}A/A(Sq^1,Sq^3)) \cong \mathrm{Ext}_{A_1}^1(\Sigma^3 \mathbb{F}_2, \Sigma^{-2}A/A(Sq^1,Sq^3))$$

And

$$\mathrm{Ext}^1_A(\Sigma^3 A//A_1, \Sigma^{-2}C/A(Sq^1, Sq^3)) \cong \mathrm{Ext}^1_{A_1}(\Sigma^3 \mathbb{F}_2, \Sigma^{-2}C/A(Sq^1, Sq^3)).$$

An A_1-module extension of $\Sigma^{-2}A/A(Sq^1,Sq^3)$ by $\Sigma^3 \mathbb{F}_2$ is determined by the values of Sq^1 and Sq^2 on the non-zero element of $\Sigma^3 \mathbb{F}_2$, and these are connected by the Adem relation $Sq^1 Sq^2 Sq^1 = Sq^2 Sq^2$.

By inspection of $\Sigma^{-2}A/A$ (Sq^1, Sq^3) and $\Sigma^{-2}C/A(Sq^1, Sq^3)$ as A_1-modules, there are precisely two such A_1-module extensions in both cases; one trivial (split) and one non-trivial (not split). Furthermore, the map of extensions induced by (4.1) induces an isomorphism

$$\mathbb{Z}/2 \cong \mathrm{Ext}^1_{A_1}(\Sigma^3 \mathbb{F}_2, \Sigma^{-2}C/A(Sq^1, Sq^3))$$

$$\xrightarrow{\cong} \mathrm{Ext}^1_{A_1}(\Sigma^3 \mathbb{F}_2, \Sigma^{-2}A/A(Sq^1, Sq^3)) \cong \mathbb{Z}/2.$$

Thus to prove that each extension is the unique non-trivial extension of its kind, it suffices to show that $H^*(Wh^{Diff}(*))$ does not split as the sum of $\Sigma^3 A/A(Sq^1, Sq^2)$ and $\Sigma^{-2}C/A(Sq^1, Sq^3)$.

Now $\Sigma^{-2}C/A(Sq^1, Sq^3)$ is 3-connected, and by Bökstedt and Waldhausen [11, 1.3] the bottom homotopy group of $Wh^{Diff}(*)$ is $\pi_3(Wh^{Diff}(*)) \cong \mathbb{Z}/2$. Hence, there is a non-trivial Sq^1 acting on the non-zero class in $H^3(Wh^{Diff}(*))$, which tells us that $\Sigma^3 A/A(Sq^1, Sq^2)$ does not split off from $H^*(Wh^{Diff}(*))$.

Remark 4.7

By (4.6), we see that the lifted cyclotomic trace map

$$\widetilde{trc} : Wh^{Diff}(*) \to \widetilde{TC}(*) \simeq \Sigma\mathbb{C}P^\infty_{-1}$$

Induces the zero homomorphism on mod 2 spectrum cohomology. The map is, nevertheless, very useful.

4.8. Question

The map ε lives in the group

$$[\mathbb{C}P_{-1}^{\infty}, \Sigma^{-2}ku] \cong [\mathbb{C}P_{+}^{\infty}, ku] = KU^{0}(\mathbb{C}P^{\infty}) \cong \mathbb{Z}[[\gamma^{1}]],$$

Where the first isomorphism is the Thom isomorphism in complex topological K-theory for the virtual complex bundle $-\gamma^{1}$ over $\mathbb{C}P_{+}^{\infty}$. To which power series in γ^{1} does ε correspond?

Proposition 4.9

The linearization map $L : TC(*) \to TC(\mathbb{Z})$ and the suspended map $\Sigma\varepsilon : \Sigma\mathbb{C}P_{-1}^{\infty} \to \Sigma^{-1}ku$ induce the same homomorphisms up to 2-adic units, on homotopy groups modulo torsion in dimensions $* = 3 \bmod 4$.

Proof

The suspended map $\Sigma\varepsilon$ is the composite

$$\Sigma\mathbb{C}P_{-1}^{\infty} \to TC(*) \xrightarrow{L} TC(\mathbb{Z}) \to \Sigma\, \mathrm{hofib}(\mathrm{trc}) \to \Sigma^{-1}ku.$$

The first map splits $TC(*)$, the second is the linearization map, the third is the connecting map in the cofiber sequence generated by $\mathrm{trc}_{\mathbb{Z}}$, and the fourth suspends a map that appears in Theorem 3.13. The first map induces an isomorphism on homotopy groups modulo torsion in all positive dimensions, since the other summand $\Sigma^{\infty}S^{0}$ has finite homotopy groups in positive dimensions. The third and fourth maps also induce an isomorphism on homotopy groups modulo torsion in dimensions $* = 3 \bmod 4$, by the calculation of $\mathrm{trc}_{\mathbb{Z}}$ in [45, 9.1], and the description of δ in Theorem 3.13.

TWO-PRIMARY HOMOTOPY OF $\mathrm{Wh}^{\mathrm{Diff}}(*)$

Let ι_{-2} be the generator in dimension -2 of $H^*(\Sigma^{-2}ku) \cong \Sigma^{-2}A/A(Sq^1, Sq^3)$, and let ι_3 be the generator in dimension 3 of $H^*(\Sigma^3 ko) \cong \Sigma^3 A/A(Sq^1, Sq^2)$.

By Proposition 4.2 the map $\varepsilon : \Sigma \mathbb{C}P_{-1}^\infty \to \Sigma^{-2}ku$ of (4.1) induces a surjection on cohomology, and we regard

$$\ker(\varepsilon^*) = \Sigma^{-2}C/A(Sq^1, Sq^3) \subset \Sigma^{-2}A/A(Sq^1, Sq^3) = H^*(\Sigma^{-2}ku)$$

as a submodule of $H^*(\Sigma^{-2}ku)$. It is thus spanned by suitable monomials $Sq^I \iota_{-2}$ taken modulo $A(Sq^1, Sq^3)\iota_{-2}$. By inspection $\ker(\varepsilon^*)$ is 3-connected. The bottom cofiber sequence in (4.1) induces the non-trivial extension

$$0 \to \ker(\varepsilon^*) \to H^*(\mathrm{Wh}^{\mathrm{Diff}}(*)) \to H^*(\Sigma^3 ko) \to 0.$$

We let $i_3 \in H^3(\mathrm{Wh}^{\mathrm{Diff}}(*))$ denote the unique lift of $\iota_3 \in H^3(\Sigma^3 ko)$. With these notations we list a basis for $H^*(\mathrm{Wh}^{\mathrm{Diff}}(*))$ in dimensions $* \leq 14$ in Table 5, together with generators for the A-module structure.

We now consider the Adams spectral sequences associated with the spectra in the cofiber sequence of spectra

$$\mathbb{C}P_{-1}^\infty \xrightarrow{i} \mathrm{hofib}(\mathrm{trc}) \xrightarrow{j} \mathrm{Wh}^{\mathrm{Diff}}(*) \tag{5.1}$$

appearing vertically in (4.1). They are

$$_cE_2^{s,t} = \mathrm{Ext}_A^{s,t}(H^*(\mathbb{C}P_{-1}^\infty), \mathbb{F}_2) \Rightarrow \pi_{t-s}(\mathbb{C}P_{-1}^\infty)_2^\wedge, \tag{5.2}$$

$$_fE_2^{s,t} = \mathrm{Ext}_A^{s,t}(H^*(\mathrm{hofib}(\mathrm{trc})), \mathbb{F}_2) \Rightarrow \pi_{t-s}(\mathrm{hofib}(\mathrm{trc}))_2^\wedge, \tag{5.3}$$

$$_wE_2^{s,t} = \mathrm{Ext}_A^{s,t}(H^*(\mathrm{Wh}^{\mathrm{Diff}}(*)), \mathbb{F}_2) \Rightarrow \pi_{t-s}(\mathrm{Wh}^{\mathrm{Diff}}(*))_2^\wedge. \tag{5.4}$$

The prefix 'c' refers to the truncated complex projective space, 'f' refers to the homotopy fiber of the cyclotomic trace map, and 'w' refers to the Whitehead spectrum. The spectral sequence $_cE_*$ was already studied in (2.15), Table 3 and Proposition 2.18, while the spectral sequence $_fE_*$ appeared in (3.17) and Table 4. The spectral sequence $_wE_*$ is displayed below, in Table 6a and Table 6b.

Diagram (5.1) induces a short exact sequence of A-modules in cohomology, by (4.6), and thus a long exact sequence of Ext-groups

$$\cdots \to {}_cE_2^{s,t} \xrightarrow{i_*} {}_fE_2^{s,t} \xrightarrow{j_*} {}_wE_2^{s,t} \xrightarrow{\partial} {}_cE_2^{s+1,t} \to \cdots. \tag{5.5}$$

By the geometric boundary theorem [40, 2.3.4], the connecting map ∂ is induced by the spectrum map $\widetilde{trc} : Wh^{Diff}(*) \to \Sigma CP_{-1}^\infty$ extending (5.1), and so each map in the long exact sequence is part of a map of spectral sequences. Furthermore, these maps are compatible with the maps in the long exact sequence in 2-completed homotopy induced by (5.1).

The E_2-term of the Adams spectral sequence (5.4) for . is displayed in dimensions $t-s \le 21$ in Table 6a and Table 6b. This was obtained by hand from a minimal resolution of $H^*(Wh^{Diff}(*))$ in internal degrees $t \le 14$, using Table 5, and using Bruner's Ext-calculator program [12] in higher dimensions. The notation in these tables is that the maps in (5.5) take a class denoted '•' in one spectral sequence to a class denoted '∘' in the following spectral sequence, i.e., $• \mapsto ∘$. As before, the classes hit by or supporting differentials are shaded gray. In bidegree (2,14) a d_2-differential hits the sum of the two given generators, which are therefore both half-shaded.

Proposition 5.6

The map $i : \Sigma CP_{-1}^\infty \to hofib(trc)$ induces a map

$$i_* : {}_cE_2^{s,t} \longrightarrow {}_fE_2^{s,t}$$

Of AdamsE_2-terms, which is surjective in dimensions$t-s \leq 2$, $t-s=4$ and$t-s \equiv 5,6 \bmod 8$. In positive dimensions $t-s \equiv 3 \bmod 8$ its image equals the three h_0-divisible classes. In other dimensions the map is zero.

Proof

Note that fE_2 has dimension 0 or 1 in each bidegree. In the range of bidegrees displayed in Tables 3, 4 and 6, the claim follows by a dimension count using exactness in (5.5). Since hofib(trc) agrees with its Bousfield K-localization in dimensions $* \geq 1$ by Theorem 3.13, the result is propagated to higher dimensions by applying suitable periodicity operators.

Proposition 5.7

In the Adams spectral sequence $_wE_*$ the non-zero differentials landing in homotopical dimension ≤ 21 are

(i)	$d_2^{s,s+8} \neq 0$ for $s \geq 0$
(ii)	$d_2^{s,s+8} \neq 0$ for $s \geq 1$, with image divisible by h_0^{s+2}.
(iii)	$d_2^{s,s+13} \neq 0$ for $0 \leq s \leq 3$. The image of $d_2^{0,13}$ contains $h_0.x + h_1.y$ for non-zero classesx,y. The image of $d_2^{s,13+s}$ for $1 \leq s \leq 3$ is divisible by h_0^{s+1}.
(iv)	$d_2^{1,15} \neq 0$ has image divisible by h_1^2.
(v)	$d_2^{1,16} \neq 0$ has image divisible by h_0.
(vi)	$d_2^{s,s+16} \neq 0$ for $s \geq 1$ is zero on the h_0 -torsion classes.
(vii)	$d_2^{s,s+18} \neq 0$ for $s=0,1$.

(viii)	$d_3^{2,20} \neq 0$ is zero on the h_1-divisible classes.
(ix)	$d_2^{s,s+19} \neq 0$ for s=3,4 is zero on the h_1-divisible classes, and takes h_0-torsion values.
(x)	$d_2^{s,s+19} \neq 0$ for s≥5, with image divisible by h_0^{s+2}.
(xi)	$d_2^{3,24} \neq 0,\ d_2^{4,25} \neq 0,\ d_2^{5,26} \neq 0,\ d_2^{6,27} \neq 0,\ d_2^{7,28} \neq 0,\ d_2^{0,22}$ $\neq 0$ and $d_2^{5,27} \neq 0$ all have rank 1.

Proof

The differentials in $_f E_*$ given in Table 4a and Table 4b induce differentials in $_w E_*$ by naturality with respect to the spectral sequence map j_* in (5.5). Likewise the differentials in $_c E_*$ given in Proposition 2.18 lift by the connecting map ∂ in (5.5) to detect differentials in $_w E_*$. Taking the h_0-multiplications in $_w E_2$ into account, this gives rise to all the differentials listed above.

It remains to check that there are no further differentials in $_w E_*$. Any such would have to map from classes '•' detected by ∂ to classes '∘' in the image of j_*. For bidegree reasons the only possible targets are the h_1-divisible classes '∘' in bidegree (s,t)=(4k,12k+3) with k≥1. These classes represent the image of $_f\pi_8 k_{+3}(\text{hofib}(\text{trc}))$ in $\pi_{8k+3}(\text{Wh}^{\text{Diff}}(*))$. Now the generator of $\pi_{8k+3}(\text{hofib}(\text{trc})) \cong \mathbb{Z}/16$ maps to the order 2 class $\eta^2 \mu_8 k_{+1}$ in $\pi_{8k+3}(\mathbb{Z}) \cong \mathbb{Z}/16$, which generates the kernel of the cyclotomic trace map $\text{trc}_{\mathbb{Z}}$ to $\pi_{8k+3}(\text{TC}(\mathbb{Z})) \cong \mathbb{Z} \oplus \mathbb{Z}/16$ by Rognes [45, 9.1]. Hence by the diagram in Theorem 1.11, the image of $\pi_8 k_{+3}(\text{hofib}(\text{trc}))$ in $\pi_{8k+3}(\text{Wh}^{\text{Diff}}(*))$ is nontrivial, and so the cited class in bidegree (4k,12k+3) must survive to the E_∞-term. Hence it is not hit by a differential.

Table 7: $H^*(K(\mathbb{Z}))$ in dimension $s \leq 14$

dim	x	$Sq^1(x)$	$Sq^2(x)$	$Sq^4(x)$	$Sq^8(x)$
0	ι_0	0	0	$Sq^4\iota_0$	$Sq^8\iota_0$
1					
2					
3	ι_3	$Sq^4\iota_0$	$Sq^2\iota_3$	$Sq^4\iota_3$	$Sq^8\iota_3$
4	$Sq^4\iota_0$	0	$Sq^6\iota_0$	0	$Sq^8Sq^4\iota_0$
5	$Sq^2\iota_3$	0	$Sq^7\iota_0$	$Sq^4Sq^2\iota_3$	$Sq^8Sq^2\iota_3$
6	$Sq^6\iota_0$	$Sq^7\iota_0$	0	$Sq^{10}\iota_0$	$Sq^{10}Sq^4\iota_0$
7	$Sq^7\iota_0$	0	0	$Sq^{11}\iota_0$	
	$Sq^4\iota_3$	0	$Sq^6\iota_3$	$Sq^6Sq^2\iota_3$	
8	$Sq^8\iota_0$	0	$Sq^{10}\iota_0$	$Sq^{12}\iota_0$	
9	$Sq^6\iota_3$	$Sq^7\iota_3$	0	$Sq^{10}\iota_3$ $+Sq^8Sq^2\iota_3$	
	$Sq^4Sq^2\iota_3$	0	$Sq^6Sq^2\iota_3$	$Sq^{13}\iota_0$	
10	$Sq^{10}\iota_0$	$Sq^{11}\iota_0$	0	0	
	$Sq^7\iota_3$	0	$Sq^8Sq^4\iota_0$ $+Sq^9\iota_3$	$Sq^{11}\iota_3$	
11	$Sq^{11}\iota_0$	0	$Sq^{13}\iota_0$		
	$Sq^8\iota_3$	$Sq^9\iota_3$	$Sq^{10}\iota_3$		
	$Sq^6Sq^2\iota_3$	$Sq^8Sq^4\iota_0$ $+Sq^9\iota_3$	0		
12	$Sq^{12}\iota_0$	$Sq^{13}\iota_0$	$Sq^{14}\iota_0$		
	$Sq^8Sq^4\iota_0$	0	$Sq^{10}Sq^4\iota_0$		
	$Sq^9\iota_3$	0	$Sq^{10}Sq^4\iota_0$		
	$\equiv Sq^7Sq^2\iota_3$				
13	$Sq^{13}\iota_0$	0			
	$Sq^{10}\iota_3$	$Sq^{11}\iota_3$			
	$Sq^8Sq^2\iota_3$	0			
14	$Sq^{14}\iota_0$				
	$Sq^{10}Sq^4\iota_0$				
	$Sq^{11}\iota_3$				

Theorem 5.8

The 2-primary homotopy groups of $\mathrm{Wh}^{\mathrm{Diff}}(*)$ in dimensions $* \leqslant 21$ are as follows:

$$\pi_n(\mathrm{Wh}^{\mathrm{Diff}}(*)) = 0 \qquad \text{for } n \leqslant 2,$$

$$\pi_3(\mathrm{Wh}^{\mathrm{Diff}}(*)) = \mathbb{Z}/2,$$

$$\pi_4(\mathrm{Wh}^{\mathrm{Diff}}(*)) = 0,$$

$$\pi_5(\mathrm{Wh}^{\mathrm{Diff}}(*)) = \mathbb{Z},$$

$$\pi_6(\mathrm{Wh}^{\mathrm{Diff}}(*)) = 0,$$

$$\pi_7(\mathrm{Wh}^{\mathrm{Diff}}(*)) = \mathbb{Z}/2,$$

$$\pi_8(\mathrm{Wh}^{\mathrm{Diff}}(*)) = 0,$$

$$\pi_9(\mathrm{Wh}^{\mathrm{Diff}}(*)) = \mathbb{Z}/2 \oplus \mathbb{Z},$$

$$\pi_{10}(\mathrm{Wh}^{\mathrm{Diff}}(*)) = (\mathbb{Z}/2)^2 \oplus \mathbb{Z}/8,$$

$$\pi_{11}(\mathrm{Wh}^{\mathrm{Diff}}(*)) = \mathbb{Z}/2,$$

$$\pi_{12}(\mathrm{Wh}^{\mathrm{Diff}}(*)) = \mathbb{Z}/4,$$

$$\pi_{13}(\mathrm{Wh}^{\mathrm{Diff}}(*)) = \mathbb{Z},$$

$$\pi_{15}(\mathrm{Wh}^{\mathrm{Diff}}(*)) = (\mathbb{Z}/2)^2,$$

$$\pi_{16}(\mathrm{Wh}^{\mathrm{Diff}}(*)) = \mathbb{Z}/2 \oplus \mathbb{Z}/8,$$

$$\pi_{17}(\mathrm{Wh}^{\mathrm{Diff}}(*)) = (\mathbb{Z}/2)^2 \oplus \mathbb{Z},$$

$$\pi_{18}(\mathrm{Wh}^{\mathrm{Diff}}(*)) = (\mathbb{Z}/2)^3 \oplus \mathbb{Z}/32,$$

$$\pi_{19}(\mathrm{Wh}^{\mathrm{Diff}}(*)) = \mathbb{Z}/2 \rtimes \mathbb{Z}/2 \rtimes \mathbb{Z}/8 \rtimes \mathbb{Z}/2,$$

$$\pi_{20}(\mathrm{Wh}^{\mathrm{Diff}}(*)) = \#2^7,$$

$$\pi_{21}(\mathrm{Wh}^{\mathrm{Diff}}(*)) = \#2^4 \oplus \mathbb{Z}.$$

In the long exact sequence in homotopy induced by the cofiber sequence

$$\mathbb{C}P^{\infty}_{-1} \xrightarrow{i} \mathrm{hofib}(\mathrm{trc}) \xrightarrow{j} \mathrm{Wh}^{\mathrm{Diff}}(*),$$

The image of j_* is $\mathbb{Z}/2$ in dimensions $n \equiv 3 \bmod 8$ and zero otherwise, for $n \leq 21$.

Proof

This follows by inspection of the E_{∞}-term of the Adams spectral sequence for $\mathrm{Wh}^{\mathrm{Diff}}(*)$, and the long exact sequence

$$\cdots \to \pi_n(\mathbb{C}P^{\infty}_{-1}) \xrightarrow{i_*} \pi_n(\mathrm{hofib}(\mathrm{trc})) \xrightarrow{j_*} \pi_n(\mathrm{Wh}^{\mathrm{Diff}}(*)) \xrightarrow{\mathrm{trc}_*} \pi_{n-1}(\mathbb{C}P^{\infty}_{-1}) \to \cdots .$$

The long exact sequence shows that $\pi_{18}(\mathrm{Wh}^{\mathrm{Diff}}(*)) \cong \pi_{17}(\mathbb{C}P^{\infty}_{-1})$, which was found in Theorem 2.11. Next, $\pi_{19}(\mathrm{Wh}^{\mathrm{Diff}}(*))$ is an extension of the torsion in $\pi_{18}(\mathbb{C}P^{\infty}_{-1}) \cong \mathbb{Z}/2 \times \mathbb{Z}/8 \times \mathbb{Z}/2 \oplus \mathbb{Z}$ by $\mathbb{Z}/2$. Also $\pi_{20}(\mathrm{Wh}^{\mathrm{Diff}}(*))$ is the kernel of a homomorphism from $\pi_{19}(\mathbb{C}P^{\infty}_{-1}) \cong \mathbb{Z}/2 \oplus \mathbb{Z}/8 \oplus \mathbb{Z}/2 \oplus \mathbb{Z}/64$ with image $\mathbb{Z}/8$. This is some group of order 2^7, denoted $\#2^7$ in the sstatement of the theorem. Lastly $\pi_{21}(\mathrm{Wh}^{\mathrm{Diff}}(*))$ is the sum of a group of order 2^4 and an infinite cyclic group, as can be read off from the E_{∞}-term of $_w E_*$.

Regarding multiplicative structure, we have the following addendum.

Lemma 5.9

The homomorphism $\eta\#: \pi_n(\mathrm{Wh}^{\mathrm{Diff}}(*)) \to \pi_{n+1}(\mathrm{Wh}^{\mathrm{Diff}}(*))$ has image $\left(\mathbb{Z}/2\right)^2$ for n=9, image $\mathbb{Z}/2$ for n=10 and is zero for all other n≤14.

The homomorphism $v\#: \pi_n(\mathrm{Wh}^{\mathrm{Diff}}(*)) \to \pi_{n+3}(\mathrm{Wh}^{\mathrm{Diff}}(*))$ has image $\mathbb{Z}/2$ for n=7 and is zero for all other n≤14.

The homomorphism $\sigma\#: \pi_n(\mathrm{Wh}^{\mathrm{Diff}}(*)) \to \pi_{n+7}(\mathrm{Wh}^{\mathrm{Diff}}(*))$ has image $\mathbb{Z}/2$ for n=5 and is zero for all other n≤11.

Proof

The non-zero multiplications listed are all detected by nontrivial h_1, h_2, or h_3 -multiplications in the Adams spectral sequence (5.4) for $\mathrm{Wh}^{\mathrm{Diff}}(*)$. To see that there are no other non-zero multiplications in this range one can use Adams filtration arguments in this spectral sequence, combined with naturality with respect to the map $\widetilde{\mathrm{trc}}: \mathrm{Wh}^{\mathrm{Diff}}(*) \to \Sigma\mathbb{CP}^\infty_{-1}$. For example, $\pi_{14}(\mathrm{Wh}^{\mathrm{Diff}}(*))$ is detected in $\pi_{13}(\mathbb{CP}^\infty_{-1})$, but the image of $\pi_7(\mathrm{Wh}^{\mathrm{Diff}}(*))$ in $\pi_6(\mathbb{CP}^\infty_{-1})$ is divisible by v and $\sigma v = 0$, so $\sigma_\# = 0$ for n=7.

COHOMOLOGY OF K($\mathbb{Z}$) AND THE LINEARIZATION MAP

We continue to implicitly complete all spectra at 2. Bökstedt's spectrum JK($\mathbb{Z}$) is the homotopy fiber of the composite

$$ko \xrightarrow{\psi^3-1} bspin \xrightarrow{c} bsu.$$

It is also homotopy equivalent to the algebraic K-theory spectrum K($\mathbb{Z}$), by Rognes and Weibel [46] and Weibel [60]. Hence there is a diagram of horizontal and vertical cofiber sequences of spectra:

$$
\begin{array}{ccccccc}
bso & = & bso & \longrightarrow & * & \longrightarrow & \Sigma bso \\
\downarrow{\scriptstyle\eta} & & \downarrow{\scriptstyle t} & & \downarrow & & \downarrow{\scriptstyle\eta} \\
spin & \xrightarrow{\ \zeta\ } & j & \longrightarrow & ko & \xrightarrow{\ \psi^3-1\ } & bspin \\
\downarrow{\scriptstyle c} & & \downarrow{\scriptstyle i} & & \| & & \downarrow{\scriptstyle c} \\
su & \longrightarrow & K(\mathbb{Z}) & \longrightarrow & ko & \longrightarrow & bsu \\
\downarrow & & \downarrow & & \downarrow & & \downarrow \\
\Sigma bso & = & \Sigma bso & \longrightarrow & * & \longrightarrow & \Sigma^2 bso .
\end{array}
\qquad (6.1)
$$

The right hand column is a connected covering of (3.2), and the second row defines the connective real image of J spectrum j. We let $t=\zeta\circ\eta$ be the composite of the Bott map $\eta : bso \to spin$ and the connecting map $\zeta : spin \to j$.

Miller and Priddy [32] define spectra $g/o_\oplus$ and ibo as the pullbacks in the following diagram:

$$
\begin{array}{ccccc}
g/o_\oplus & \longrightarrow & ibo & \longrightarrow & S^0 \\
\downarrow & & \downarrow & & \downarrow{\scriptstyle e} \\
bso & \xrightarrow{\ \eta\ } & spin & \xrightarrow{\ \zeta\ } & j .
\end{array}
\qquad (6.2)
$$

(More precisely, they define the underlying infinite loop spaces $G/O_\oplus = \Omega^\infty g/o_\oplus$ and $IBO=\Omega^\infty ibo$.) Here $e : S^0 \to j$ is the map representing the real Adams e-invariant. Its fiber c is the cokernel of J spectrum, which is K-acyclic. Thus the unit map $i : S^0 \to K(\mathbb{Z})$ factors, uniquely up to homotopy, as e composed with $i : j \to K(\mathbb{Z})$. By (6.1) the cofiber of the lower composite map $\zeta\eta$ in (6.2) is $K(\mathbb{Z})$. Hence there is a cofiber sequence

$$g/o_\oplus \to S^0 \xrightarrow{i} K(\mathbb{Z}) \tag{6.3}$$

of 2-complete spectra. Thus there is a fiber sequence $G/O_\oplus \to QS^0 \to K(\mathbb{Z})$

of underlying infinite loop spaces, and we might write $G/O_\oplus \to IK(\mathbb{Z})$ as the 'ideal' in $QS^0 = \Omega^\infty S^0$ defining $K(\mathbb{Z})$ (at the prime 2).

We compute the mod 2 spectrum cohomology $H^*(K(\mathbb{Z}))$ by means of the cofiber sequence $su \to K(\mathbb{Z}) \to ko$, where $su \simeq \Sigma^3 ku$. In view of (6.3) this also determines $H^*(g/o_\oplus)$. Miller and Priddy conjecture in [32] that $G/O_\oplus \simeq G/O$ as infinite loop spaces. If confirmed, this would also lead to a calculation of the spectrum cohomology $H^*(g/o)$. It is known that $G/O \simeq G/O_\oplus$ as 2-complete spaces, and that $H_*(G/O;F_2) \cong H_*(G/O_\oplus;F_2)$ as Hopf algebras over the Steenrod– and Dyer–Lashof algebras, by un-published calculations of J. Tornehave.

Theorem 6.4

The mod 2 spectrum cohomology of $K(\mathbb{Z})$ is the unique nontrivial ex-tension of A-modules

$$A/A(Sq^1, Sq^2) \to H^*(K(\mathbb{Z})) \to \Sigma^3 A/A(Sq^1, Sq^3).$$

The A-module $H^*(\Sigma g/o_\oplus)$ is the connected cover of $H^*(K(\mathbb{Z}))$ i.e., the kernel of the augmentation $H^*(K(\mathbb{Z})) \to F_2$.

Proof

We use the cofiber sequence $K(\mathbb{Z}) \to ko \to bsu$ where the right hand map is the composite of $\psi^3 - 1 : ko \to bspin$ and $c : bspin \to bsu$. The induced map

$$\Sigma^4 A/A(Sq^1, Sq^3) \cong H^*(bsu) \to H^*(ko) \cong A/A(Sq^1, Sq^2)$$

Is the zero homomorphism. For the complexification map c induces multiplication by 2 on π_4, and thus the zero map on H^4. Thus the long exact sequence in spectrum cohomology decomposes as the A-module extension above. The second claim follows from the cofiber sequence

$$S^0 \to K(\mathbb{Z}) \to \Sigma g/o_\oplus.$$

It remains to characterize the extension. There are precisely two such A-module extensions, since

$$\mathrm{Ext}^1_A(\Sigma^3 A/A(Sq^1, Sq^3), A/A(Sq^1, Sq^2)) \cong \mathrm{Ext}^1_{E_1}(\Sigma^3 \mathbb{F}_2, A/A(Sq^1, Sq^2)) \cong \mathbb{Z}/2.$$

Here $E_1 \subset A$ is the exterior algebra generated by Sq^1 and $Q_1 = Sq^3 + Sq^2 Sq^1$. We know that $H^*(K(\mathbb{Z}))$ is a non-trivial extension, because $H_3^{\mathrm{spec}}(k(\mathbb{Z})); \mathbb{Z}_2 \cong \pi_2(g/o_\oplus) \cong \mathbb{Z}/2$ implies that there is a non-zero Sq^1 from dimension 3 to dimension 4 in $H^*(K(\mathbb{Z}))$.

We list a monomial basis for $H^*(K(\mathbb{Z}))$ in dimension $s \leq 14$ in Table 7. It differs from $H^*(\Sigma g/o_\oplus)$ only in dimension 0. The notation is that $i_0 \in H^0(k(\mathbb{Z}))$ pulls back from the generator of $H^0(ko)$, while $i_3 \in H^3(k(\mathbb{Z}))$ is the unique lift of the generator in dimension 3 of $H^*(su) \cong \Sigma^3 A/A(Sq^1, Sq^3)$. We have chosen $Sq^9(\iota_3) = Sq^1 Sq^8(\iota_3)$ as the lift in $H^*(K(\mathbb{Z}))$ of $Sq^9 \iota_3 = Sq^7 Sq^2 \iota_3$ in $H^*(su)$.

The linearization map $L : A(*) \to k(\mathbb{Z})$ from [53] and Corollary 1.8 is compatible with the unit maps from S^0. When combined with the pullback diagram (6.2) defining $g/o_\oplus$ it yields the following spectrum level diagram with horizontal cofiber sequences:

$$
\begin{array}{ccccc}
S^0 & \xrightarrow{\ i\ } & A(*) & \dashrightarrow & \mathrm{Wh}^{\mathrm{Diff}}(*) \\
\Big\| & & \Big\downarrow{\scriptstyle L} & & \Big\downarrow{\scriptstyle \bar{L}} \\
S^0 & \xrightarrow{\ i\ } & K(\mathbb{Z}) & \longrightarrow & \Sigma g/o_{\oplus} \\
\Big\downarrow{\scriptstyle e} & & \Big\| & & \Big\downarrow \\
j & \xrightarrow{\ i\ } & K(\mathbb{Z}) & \longrightarrow & \Sigma bso\,.
\end{array}
$$

$$(6.5)$$

Proposition 6.6

The reduced linearization map $\bar{L}: \mathrm{Wh}^{\mathrm{Diff}}(*) \to \Sigma g/o_{\oplus}$ is a rational equivalence, but induces the zero homomorphism between the bottom homotopy groups $\pi_3(\mathrm{Wh}^{\mathrm{Diff}}(*)) \cong \mathbb{Z}/2$ and $\pi_3(\Sigma g/o_{\oplus}) \cong \mathbb{Z}/2$. The induced map on mod2 spectrum cohomology

$$\bar{L}^* : H^*(\Sigma g/o_{\oplus}) \to H^*(\mathrm{Wh}^{\mathrm{Diff}}(*))$$

is zero in all dimensions.

Proof

The linearization map $L: A(*) \to K(\mathbb{Z})$ is a rational equivalence between spectra of finite type, by Waldhausen [53, 2.2], so its 2-adic completion is also a rational equivalence. Comparison with (6.5) shows that also $\bar{L}$ is a rational equivalence.

The homomorphism $\pi_3(\bar{L})$ is induced from the homomorphism

$$\pi_3(L) : \pi_3(A(*)) \cong \mathbb{Z}/24 \oplus \mathbb{Z}/2 \to K_3(\mathbb{Z}) \cong \mathbb{Z}/48$$

By passage to the quotient with respect to subgroups $\pi_3^s \cong \mathbb{Z}/24$ on both sides. Algebraically, the only possibility is that $\pi_3(\bar{L})=0$.

In cohomology we have the following map of extensions of A-modules:

$$\begin{array}{ccccc}
H^*(\Sigma g/o_\oplus) & \longrightarrow & H^*(K(\mathbb{Z})) & \xrightarrow{\ i^*\ } & \mathbb{F}_2 \\
\downarrow{\bar{L}^*} & & \downarrow{L^*} & & \| \\
H^*(\mathrm{Wh}^{\mathrm{Diff}}(*)) & \longrightarrow & H^*(A(*)) & \xrightarrow{\ i^*\ } & \mathbb{F}_2.
\end{array}$$

The lower extension is split, as in Theorem 4.5. Here $H^*(K(\mathbb{Z}))$ is generated as and A-module by classes ι_0 and ι_3, as in Theorem 6.4 and Table 7. The class ι_0 maps to the split summand F_2 of $H^*(A(*))$, hence the submodule it generates maps to zero in positive degrees. Likewise ι_3 maps to zero by the π_3-calculation above and the Hurewicz theorem. Thus L^* is zero in positive degrees, and $\bar{L}^*$ is zero in all degrees.

Corollary 6.7

There is a long exact sequence in mod2 spectrum cohomology

$$\cdots \to H^*(TC(\mathbb{Z})) \xrightarrow{L^* \oplus \mathrm{trc}_{\mathbb{Z}}^*} H^*(TC(*)) \oplus H^*(K(\mathbb{Z})) \xrightarrow{\mathrm{trc}_*^* - L^*} H^*(A(*)) \xrightarrow{\partial} \cdots.$$

Here $L : A(*) \to K(\mathbb{Z})$ and $\mathrm{trc}_* : A(*) \to TC(*)$ induce zero maps in positive dimensions, ∂ induces an injective map in positive dimensions, and $L : TC(*) \to TC(\mathbb{Z})$ and $\mathrm{trc}_{\mathbb{Z}} : K(\mathbb{Z}) \to TC(\mathbb{Z})$ both induce surjections in all dimensions.

Proof

The sequence arises by applying mod 2 spectrum cohomology to the homotopy cartesian square inCorollary 1.8 for X=*. The assertions for $L : A(*) \to K(\mathbb{Z})$ and trc_* follow from Remark 4.7 and Proposition 6.6.

The rest follows by exactness. In fact $L^* \oplus \mathrm{trc}_{\mathbb{Z}}^*$ will be surjective in positive degrees, which is stronger than the stated conclusion.

Remark 6.8

The 'rigid tube' map from [55, Section 3] provides a space level map of horizontal fiber sequences

$$
\begin{array}{ccc}
G/O & \longrightarrow & BSO \xrightarrow{\ j\ } BSG \\
\downarrow{\scriptstyle hw} & & \downarrow{\scriptstyle s} \qquad \downarrow{\scriptstyle w} \\
\Omega\mathrm{Wh}^{\mathrm{Diff}}(*) & \longrightarrow & QS^0 \xrightarrow{\ i\ } A(*).
\end{array}
$$

We call the left vertical map hw the Hatcher–Waldhausen map. It was proved in [41] that this gives a diagram of infinite loop maps if one uses a multiplicative infinite loop space structure on each of the spaces in the lower row. However, these are generally different from the additive infinite loop space structures we have been considering in this paper.

Let $\Omega\mathrm{Wh}^{\mathrm{Diff}}_{\otimes}(*)$ denote the spectrum with underlying infinite loop space given as the homotopy fiber of the unit map $i : SG = Q(S^0)_1 \to A(*)_1$ with the multiplicative infinite loop space structures.

It can be read off from Tables 5–7 that the (space level) Hatcher–Waldhausen map $hw : G/O \to \Omega\mathrm{Wh}^{\mathrm{Diff}}(*)$, $i : SG = Q(S^0)_1 \to A(*)_1$ does not admit a four-fold delooping, when the target is given the additive infinite loop space structure. For by Waldhausen [55], $\pi_2(hw) : \mathbb{Z}/2 \cong \mathbb{Z}/2$ is an isomorphism, and a k-invariant argument (see Theorem 7.5 below) shows that $\pi_4(hw) : \mathbb{Z} \to \mathbb{Z}$ is a 2-adic equivalence. If hw were to admit a four-fold delooping then $\sigma \cdot hw(x) = hw(\sigma \cdot x)$ for any $x \in \pi_4(G/O)$. But $\pi_{11}(G/O) = 0$, while the minimal resolution leading to Table 6 shows that there is a non-zero h_3-multiplication from the class representing the generator of $\pi_4(\Omega\mathrm{Wh}^{\mathrm{Diff}}(*))$ to the class representing the element of order 2 in $\pi_{11}(\Omega\mathrm{Wh}^{\mathrm{Diff}}(*))$. See also Lemma 5.9. This contradicts the existence of the four-fold delooping. Note that we did not specify a choice of four-fold delooping of G/O in this argument, so it applies to both $\Omega(\Sigma^4 g/o)$ and $\Omega^{\infty}(\Sigma^4 g/o_{\oplus})$, in case they are different.

The spectrum map $g/o \to \Omega \mathrm{Wh}_\otimes^{\mathrm{Diff}}(*)$ constructed geometrically in [41] thus shows that the spectra $\Omega \mathrm{Wh}^{\mathrm{Diff}}(*)$ and $\Omega \mathrm{Wh}_\otimes^{\mathrm{Diff}}(*)$ cannot be homotopy equivalent.

A SPECTRUM MAP FROM $\Omega \mathrm{Wh}^{\mathrm{Diff}}(*)$ TO $\Sigma G/O \oplus$

Observe by inspection of Table 5 and Table 7 that $H^*(\mathrm{Wh}^{\mathrm{Diff}}(*))$ and $H^*(\Sigma g / o_\oplus)$ are abstractly isomorphic as A-modules in dimensions $* \leq 9$. In this chapter we construct a spectrum map

$$M : \mathrm{Wh}^{\mathrm{Diff}}(*) \longrightarrow \Sigma g/o_\oplus$$

Inducing an isomorphism in these dimensions. As before, all spectra are implicitly 2-completed in this section.

Lemma 7.1

There is a spectrum map $m : \mathrm{hofib}(\mathrm{trc}) \to K(\mathbb{Z})$ making the following diagram of horizontal cofiber sequences commute:

$$
\begin{array}{ccccc}
\mathrm{hofib}(\mathrm{trc}) & \longrightarrow & \Sigma^{-2} ku & \xrightarrow{\;\delta\;} & \Sigma^4 ko \\
\downarrow{\scriptstyle m} & & \downarrow{\scriptstyle r\beta^{-1}} & & \downarrow{\scriptstyle \beta^2 \Sigma^4 c} \\
K(\mathbb{Z}) & \longrightarrow & ko & \xrightarrow{\;c(\psi^3-1)\;} & bsu\,.
\end{array}
$$

Proof

The maps in the right hand square are characterized (up to homotopy) by their K-localizations, and after K-localization we can compute

$$\beta^2 \Sigma^4 c \circ L_K \delta = \beta^2 \circ \Sigma^4 c \circ \Sigma^4 r \circ \beta^{-2} \circ (\psi^3 - 1) \circ \beta^{-1} = c(\psi^3 - 1) \circ r\beta^{-1}.$$

Hence the right hand square homotopy commutes. We let m be the induced map of horizontal homotopy fibers.

Lemma 7.2

There is a spectrum map $M : \mathrm{Wh}^{\mathrm{Diff}}(*) \to \Sigma g / o_\oplus$ making the following diagram of horizontal cofiber sequences commute:

$$
\begin{array}{ccccc}
\mathbb{C}P^\infty_{-1} & \xrightarrow{i} & \mathrm{hofib(trc)} & \xrightarrow{j} & \mathrm{Wh}^{\mathrm{Diff}}(*) \\
\downarrow & & \downarrow{m} & & \downarrow{M} \\
S^0 & \xrightarrow{i} & K(\mathbb{Z}) & \longrightarrow & \Sigma g / o_\oplus .
\end{array}
$$

Proof

We must show that the composite map

$$
\mathbb{C}P^\infty_{-1} \xrightarrow{i} \mathrm{hofib(trc)} \xrightarrow{m} K(\mathbb{Z}) \to \Sigma g / o_\oplus
$$

Is null homotopic. Consider the following diagram of horizontal and vertical cofiber sequences:

$$
\begin{array}{ccccc}
S^0 & =\!=\!=\!=\!= & S^0 & & \\
\downarrow{i} & & \downarrow{i} & & \\
K(\mathbb{Z}) & \longrightarrow & ko & \xrightarrow{c(\psi^3-1)} & bsu \\
\downarrow & & \downarrow & & \| \\
\Sigma g / o_\oplus & \longrightarrow & ko/S^0 & \longrightarrow & bsu .
\end{array}
$$

We have $[\mathbb{C}P^\infty_{-1}, su] = 0$ by an application of the Atiyah–Hirzebruch spectral sequence, so we can identify $[\mathbb{C}P^\infty_{-1}, K(\mathbb{Z})]$ with the kernel of

$$c(\psi^3 - 1)_\# : [\mathbb{C}P^\infty_{-1}, ko] \to [\mathbb{C}P^\infty_{-1}, bsu].$$

By another calculation with the Atiyah–Hirzebruch spectral sequence using [1] and [2], this kernel is isomorphic to $\mathbb{Z}_2$, and is generated by the composite map

$$\mathbb{C}P^\infty_{-1} \to \mathbb{C}P^\infty_+ \to S^0 \overset{i}{\to} K(\mathbb{Z}).$$

The left hand map pinches the bottom cell to a point; the middle map retracts $\mathbb{C}P^\infty$ to a point. The composite maps to zero in $[\mathbb{C}P^\infty_{-1}, \Sigma g / o_\oplus]$, so m extends to a map M as claimed.

Lemma 7.3

The map $M : \mathrm{Wh}^{\mathrm{Diff}}(*) \to \Sigma g / o_\oplus$ induces an isomorphism on π_3.

Proof

Consider the maps of long exact sequences of homotopy groups induced by the diagrams in Lemma 7.1 and Lemma 7.2. The isomorphism $\pi_4(\beta^2 \Sigma^4 c) : \mathbb{Z} \cong \mathbb{Z}$ passes to quotient isomorphisms $\pi_3(m) : \mathbb{Z}/16 \cong \mathbb{Z}/16$ and $\pi_3(M) : \mathbb{Z}/2 \cong \mathbb{Z}/2$.

Theorem 7.4

There is a spectrum map

$$M : \mathrm{Wh}^{\mathrm{Diff}}(*) \to \Sigma g/o_\oplus$$

Inducing an isomorphism on mod2 spectrum cohomology in dimensions $* \leq 9$. SoMis precisely 9-connected, and induces a map of spaces

$$\Omega M : \Omega \mathrm{Wh}^{\mathrm{Diff}}(*) \to G/O_\oplus \simeq G/O$$

Such that $\pi_*(\Omega M)$ is an isomorphism for $* \leq 8$.

Proof

The A-module homomorphism $M^* : H^*(\Sigma g / o_\oplus) \to H^*(\mathrm{Wh}^{\mathrm{Diff}}(*))$ is an isomorphism in degree 3 by Lemma 7.3. We can then compute M^* in dimensions $* \leq 14$ from Table 5 and Table 7, finding that $H^*(\mathrm{hofib}(M))$ is 9-connected, has rank 1 in each dimension $10 \geq * \leq 13$, and has rank ≥ 1 in dimension 14. Thus ΩM is 8-connected, and the surjection $\pi_8(\Omega M)$ is in fact an isomorphism, since both its source and target are isomorphic to $\mathbb{Z} \oplus \mathbb{Z}/2$.

Theorem 7.5

The Hatcher–Waldhausen map $\mathrm{hw} : G/o \to \Omega\mathrm{Wh}^{\mathrm{Diff}}(*)$ induces an isomorphism on 2-primary homotopy in dimensions $* \leq 8$, and an injection on 2-primary homotopy in dimensions $* \leq 13$. Its 2-completion is thus precisely 8-connected.

Proof

Let PnX denote the nth Postnikov section of a (simple) space X. The map

$$P^2(hw) : P^2 G/O \to P^2 \Omega\mathrm{Wh}^{\mathrm{Diff}}(*)$$

is a homotopy equivalence by Lemma 7.3. The k-invariants of G/O and $\Omega\mathrm{Wh}^{\mathrm{Diff}}(*)$ all lift to spectrum cohomology, since these are infinite loop spaces, and are abstractly isomorphic for $n \leq 8$ by Theorem 7.4. They can be partly read off from the minimal resolution for $H^*(\mathrm{Wh}^{\mathrm{Diff}}(*))$ that was used to generate Table 6, yielding the following facts: Let $\beta_1 : K(\mathbb{Z}/2, n) \to K(\mathbb{Z}, n+1)$ be the mod2 Bockstein map, and let $i_1 : K(\mathbb{Z}, n) \to K(\mathbb{Z}/2, n)$ be the mod2 reduction map. Then $i_1 \beta_1 = \mathrm{Sq}^1$.

Form n let pnm:PmX→PnX be a projection in the Postnikov system. Then

$$k^5 : K(\mathbb{Z}/2, 2) \simeq P^2 \Omega \mathrm{Wh}^{\mathrm{Diff}}(*) \to K(\mathbb{Z}, 5)$$

is $\beta_1 Sq^2$, while

$$k^7 : P^4 \Omega \mathrm{Wh}^{\mathrm{Diff}}(*) \to K(\mathbb{Z}/2, 7)$$

factors as $Sq^5 p_2{}^4$. The last k-invariant we consider is

$$k^9 = k_1^9 \times k_2^9 : P^6 \Omega \mathrm{Wh}^{\mathrm{Diff}}(*) \to K(\mathbb{Z}/2 \oplus \mathbb{Z}, 9) \simeq K(\mathbb{Z}/2, 9) \times K(\mathbb{Z}, 9).$$

Its projection $k_2{}^9$ to $K(\mathbb{Z}, 9)$ factors over $p_4{}^6$, and the composite

$$K(\mathbb{Z}, 4) \to P^4 \Omega \mathrm{Wh}^{\mathrm{Diff}}(*) \xrightarrow{\bar{k}_2^9} K(\mathbb{Z}, 9)$$

Is $\beta_1 Sq^4 i_1$. Here $k_2^9 = k_2^{-9} o p_4^6$.

Considering the maps of Postnikov sections $P^n(hw) : P^n G/O \to P^n \Omega \mathrm{Wh}^{\mathrm{Diff}}(*)$ and comparing the k-invariants for G/O and $\Omega \mathrm{Wh}^{\mathrm{Diff}}(*)$, it follows that also P^4 (hw) and P^6(hw) are homotopy equivalences, and that P^8(hw) induces an isomorphism on π_8 modulo the torsion subgroups. Hence π_*(hw) is an isomorphism for $* \leq 7$. In particular the image of $v^2 \in \pi_6(SG) \cong \pi_6 S$ in $\pi_6(G/O)$ maps to the generator of $\pi_6 \Omega \mathrm{Wh}^{\mathrm{Diff}}(*)$.

The 2-torsion in $\pi_8(G/O)$ is the image of $\bar{v} \in \pi_8(SG) \cong \pi_8^S$, satisfying $\eta.\bar{v} = v.v^2$. The image of $\bar{v}$ in $\pi_8(\Omega \mathrm{Wh}^{\mathrm{Diff}}(*))$ is non-zero, because $\eta.hw(\bar{v}) = v.hw(v^2)$ is non-zero, as can be seen from Table 6a or detected by ΩM. Hence π_8(hw) is also an isomorphism on the torsion in dimension 8. So hw is 8-connected, but cannot be 9-connected because $\pi_9(G/O) = (\mathbb{Z}/2)^2$ cannot surject to $\pi_9(\Omega Wh^{Diff}(*)) = (\mathbb{Z}/2)^2 \oplus \mathbb{Z}/8$.

The non-zero multiplications by η in $\pi_*(\Omega Wh^{Diff}(*))$ given in Lemma 5.9 then imply that $\pi n(hw)$ is injective for $9 \leq n \geq 11$ and $n=13$. Finally, $\pi_{12}(hw)$ is injective since $\pi_{12}(G/O) = \mathbb{Z}$ and hw is a rational equivalence [5].

ACKNOWLEDGEMENTS

The main part of this work was done in December 1997 during visits to Aarhus and Bielefeld. The author thanks M. Bökstedt, I. Madsen, J. Tornehave and F. Waldhausen for helpful discussions and hospitality.

REFERENCES

1. J.F. Adams, Vector .elds on spheres, Ann. Math. 75 (1962) 603–632.
2. J.F. Adams, G. Walker, On complex Stiefel manifolds, Proc. Cambridge Philos. Soc. 61 (1965) 81–103.
3. M.F. Atiyah, Thom complexes, Proc. London Math. Soc. 11 (1961) 291–310.
4. S. Bloch, S. Lichtenbaum, A spectral sequence for motivic cohomology, Invent. Math., to appear.
5. M. B Pokstedt, The rational homotopy type of YWhDi1 (∗), in: I. Madsen, B. Oliver (Eds.), Algebraic topology, Proceedings of a conference, Aarhus 1982, Lecture Notes in Mathematics, Vol. 1051, Springer, 1984, pp. 25–37.
6. M. B Pokstedt, Topological Hochschild homology, preprint, Bielefeld, Topology, to appear.
7. M. B Pokstedt, G. Carlsson, R. Cohen, T. Goodwillie, W.-C. Hsiang, I. Madsen, On the algebraic K-theory of simply connected spaces, Duke Math. J. 84 (1996) 541–563.
8. M. B Pokstedt, W.-C. Hsiang, I. Madsen, The cyclotomic trace and algebraic K-theory of spaces, Invent. Math. 111 (1993) 865–940.
9. M. B Pokstedt, I. Madsen, Topological cyclic homology of the integers, Asterisque 226 (1994) 57–143.
10. M. B Pokstedt, I. Madsen, Algebraic K-theory of local number .elds: the unrami. ed case, Prospects in Topology, Princeton, NJ, 1994, Annals of Mathematics Studies, Vol. 138, Princeton University Press, 1995, pp. 28–57.

11. M. B Pokstedt, F. Waldhausen, The map BSG → A(∗) → QS0 , Algebraic topology and algebraic K-theory, Proceedings of a Conference, Princeton/New Jersey, 1983. Annals of Mathematics Studies, Vol. 113, Princeton University Press, 1987, pp. 418– 431.

12. R.R. Bruner, Ext-calculator, Computer code available from rrb@math.wayne.edu.

13. D. Burghelea, R. Lashof, The homotopy type of the space of di1eomorphisms. I, Trans. Amer. Math. Soc. 196 (1974) 1–36.

14. B.I. Dundas, Relative K-theory and topological cyclic homology, Acta Math. 179 (1997) 223–242.

15. B.I. Dundas, The cyclotomic trace for symmetric monoidal categories, in: K. Grove, I.H. Madsen, E.K. Pedersen (Eds.), Geometry and Topology: Aarhus, Contemporary Mathematics, Vol. 258, AMS, Providence, RI, 2000, pp. 121–144.

16. B.I. Dundas, R. McCarthy, Topological Hochschild homology of ring functors and exact categories, J. Pure Appl. Algebra 109 (1996) 231–294.

17. W.G. Dwyer, Twisted homological stability for general linear groups, Ann. of Math. 111 (1980) 239–251.

18. W.G. Dwyer, S.A. Mitchell, On the K-theory spectrum of a ring of algebraic integers, K-Theory 14 (1998) 201–263.

19. A.D. Elmendorf, I. Kriz, M.A. Mandell, J.P. May, Rings, Modules, and Algebras in Stable Homotopy Theory, Mathematical Surveys and Monographs, Vol. 47, AMS, Providence, RI, 1997.

20. T.G. Goodwillie, Relative algebraic K-theory and cyclic homology, Ann. of Math. 124 (1986) 347–402.

21. A.E. Hatcher, Concordance spaces, higher simple-homotopy theory, and applications, Algebraic and geometric Topology, Stanford=California, 1976, Proceedings of Symposia in Pure Mathematics, Vol. 32, Part 1, 1978, pp. 3–21.

22. L. Hesselholt, I. Madsen, On the K-theory of .nite algebras over Witt vectors of perfect .elds, Topology 36 (1997) 29–101.

23. K. Igusa, The stability theorem for smooth pseudoisotopies, K-Theory 2 (1988) 1–355.

24. R.C. Kirby, L.C. Siebenmann, Foundational Essays on Topological Manifolds, Smoothings and Triangulations, Annals of Mathematics Studies, Vol. 88, Princeton University Press, Princeton, NJ, 1977.

25. K. Knapp, Some Applications of K-Theory to Framed Bordism: e-Invariant and Transfer, Habilitationsschrift, Bonn, 1979.

26. L.G. Lewis Jr., J.P. May, M. Steinberger, Equivariant Stable Homotopy Theory, Lecture Notes in Mathematics, Vol. 1213, Springer, Berlin, 1986.

27. M. Lydakis, Smash products and Z-spaces, Math. Proc. Cambridge Philos. Soc. 126 (1999) 311–328.

28. I. Madsen, Algebraic K-theory and traces, Current Developments in Mathematics, International Press, Cambridge, MA, 1995, pp. 191–321.

29. I. Madsen, V. Snaith, J. Tornehave, In.nite loop maps in geometric topology, Math. Proc. Cambridge Phil. Soc. 81 (1977) 399–429.

30. J.P. May, F. Quinn, N. Ray, with contributions by J. Tornehave, E∞ Ring Spaces and E∞ Ring Spectra, Lecture Notes in Mathematics, Vol. 577, Springer, Berlin, 1977.

31. R. McCarthy, Relative algebraic K-theory and topological cyclic homology, Acta Math. 179 (1997) 197–222.

32. H.R. Miller, S.B. Priddy, On G and the stable Adams conjecture, Geom. Appl. Homotopy Theory II, Proceedings of Conference, Evanston, 1977, Lecture Notes in Mathematics, Vol. 658, Springer, Berlin, 1978, pp. 331–348.

33. M. Mimura, H. Toda, The (n + 20)-th homotopy groups of n-spheres, J. Math. Kyoto Univ. 3 (1963) 37–58.

34. R.E. Mosher, Some stable homotopy of complex projective space, Topology 7 (1968) 179–193.

35. J. Mukai, The S1 -transfer map and homotopy groups of suspended complex projective spaces, Math. J. Okayama Univ. 24 (1982) 179–200.

36. J. Mukai, On stable homotopy of the complex projective space, Japan J. Math. 19 (1993) 191–216.

37. J. Mukai, The element UA is not in the image of the S1 -transfer, Math. J. Okayama Univ. 36 (1994) 133–143.

38. D. Quillen, On the cohomology and K-theory of the general linear groups over a .nite .eld, Ann. of Math. 96 (1972) 552–586.

39. D. Quillen, Higher algebraic K-theory. I, Algebraic K-Theory I, Proceedings of Conference in Battelle Inst. 1972, Lecture Notes Mathematics, Vol. 341, Springer, Berlin, 1973, pp. 85 –147.

40. D.C. Ravenel, Complex Cobordism and Stable Homotopy Groups of Spheres, Pure and Applied Mathematics, Vol. 121, Academic Press, 1986.

41. J. Rognes, The Hatcher–Waldhausen map is a spectrum map, Math. Ann. 299 (1994) 529–549.

42. J. Rognes, after M. B Pokstedt, Trace maps from the algebraic K-theory of the integers, J. Pure Appl. Algebra 125 (1998) 277–286.

43. J. Rognes, The product on topological Hochschild homology of the integers with mod four coeQcients, J. Pure Appl. Algebra 134 (1999) 210–217.

44. J. Rognes, Topological cyclic homology of the integers at two, J. Pure Appl. Algebra 134 (1999) 219–286.

45. J. Rognes, Algebraic K-theory of the two-adic integers, J. Pure Appl. Algebra 134 (1999) 287–326.

46. J. Rognes, C. Weibel, Two–primary algebraic K-theory of rings of integers in number .elds, J. Amer. Math. Soc. 13 (2000) 1–54.

47. G. Segal, Categories and cohomology theories, Topology 13 (1974) 293–312.

48. A.A. Suslin, On the K-theory of algebraically closed .elds, Invent. Math. 73 (1983) 241–245.

49. A.A. Suslin, Algebraic K-theory and motivic cohomology, Proceedings of the International Congress of Mathematicians, Zurich, 1994, BirkhP P auser, 1995, pp. 342–351.

50. H. Toda, Composition Methods in Homotopy Groups of Spheres, Annals of Mathematics Studies, Vol. 49, Princeton University Press, Princeton, NJ, 1962.

51. V. Voevodsky, The Milnor Conjecture, preprint, K-Theory Archive, no. 170, 1996.

52. W. Vogell, The involution in the algebraic K-theory of spaces, Algebraic and geometric topology, Proceedings of a Conference, New Brunswick=USA, 1983, Lecture Notes in Mathematics, Vol. 1126, Springer, Berlin, 1985, pp. 277–317.

53. F. Waldhausen, Algebraic K-theory of topological spaces. I, Algebraic and geometric Topology, Stanford= California 1976, Proceedings of Symposia in Pure Mathematics, Vol. 22, Part 1, 1978, pp. 35 – 60.

54. F. Waldhausen, Algebraic K-theory of topological spaces. II. Algebraic topology, Proceedings of a Symposium, Aarhus, 1978, Lecture Notes in Mathematics, Vol. 763, Springer, Berlin, 1979, pp. 356 –394.

55. F. Waldhausen, Algebraic K-theory of spaces, a manifold approach, Current trends in algebraic topology, Seminar, London=Ontario, 1981, CMS Conference Proceedings 2, Part 1, 1982, pp. 141–184.

56. F. Waldhausen, Algebraic K-theory of spaces, localization, and the chromatic .ltration of stable homotopy, Algebraic topology, Proceedings of a Conference, Aarhus, 1982, Lecture Notes in Mathematics, Vol. 1051, Springer, Berlin, 1984, pp. 173–195.

57. F. Waldhausen, Algebraic K-theory of spaces, Algebraic and geometric topology, Proceedings of the Conference, New Brunswick=USA, 1983, Lecture Notes in Mathematics, Vol. 1126, Springer, Berlin, 1985, pp. 318– 419.

58. F. Waldhausen, Algebraic K-theory of spaces, concordance, and stable homotopy theory, Algebraic topology and algebraic K-theory, Proceedings of the Conference, Princeton, NJ (USA), Annals of Mathematics Studies, Vol. 113, 1987, pp. 392–417.

59. F. Waldhausen, On the construction of the Kan loop group, Doc. Math. 1 (1996) 121–126.

60. C. Weibel, The 2-torsion in the K-theory of the integers, C. R. Acad. Sci. Paris 324 (1997) 615–620.

61. M. Weiss, B. Williams, Automorphisms of manifolds and algebraic K-theory. I, K-Theory 1 (1988) 575–626.

62. M. Weiss, B. Williams, Assembly, Novikov conjectures, Index theorems and Rigidity, Cambridge University Press, Cambridge, 1995, pp. 332–352.

63. M. Weiss, B. Williams, Automorphisms of manifolds, in: S. Cappell, A. Ranicki, J. Rosenberg (Eds.), Surveys on Surgery Theory: Vol. 2, Annals of Mathematics Studies, Vol. 149, Princeton University Press, Princeton, NJ, 2001, pp. 165–220.

CITATION

1. John Rognes, Two-primary algebraic K-theory of pointed spaces, Topology, Volume 41, Issue 5, September 2002, Pages 873-926, ISSN 0040-9383, http://dx.doi.org/10.1016/S0040-9383(01)00005-2.

Probability Theory Predicts that Chunking into groups of three or four items Increases the Short-term Memory Capacity

Motohisa Osaka

Department of Basic Science, Nippon
Veterinary and Life Science University,
Tokyo, Japan

ABSTRACT

Short-term memory allows individuals to recall stimuli, such as numbers or words, for several seconds to several minutes without rehearsal. Although the capacity of short-term memory is considered to be 7 ± 2 items, this can be increased through a process called chunking. For example, in Japan, 11-digit cellular phone numbers and 10-digit toll free numbers are chunked into three groups of three or four digits: 090-XXXX-XXXX and 0120-XXX-XXX, respectively. We use probability theory to predict that the most effective chunking involves groups of three or four items, such as in phone numbers. However, a 16-digit credit card number exceeds the capacity of short-term memory, even when chunked into groups of four digits, such as XXXX-XXXX-XXXX-XXXX. Based on these data, 16-digit credit card numbers should be sufficient for security purposes.

INTRODUCTION

Short-term memory allows stimuli, such as numbers or words, to be recalled for several seconds to several minutes without rehearsal. Miller (1956) reported that the storage capacity of short-term memory was 7

± 2 items, naming this "the magical number" [1]. He concluded that human "channel capacity" does not exceed a few bits and that unambiguous judgment of one-dimensional stimuli (i.e., all numbers) can be made from 7 ± 2 categories. Recently, Cowan (2001) reported that the capacity of short-term memory is 4 - 5 items [2]. Baddeley (1994) thought highly of "magical number seven", saying that it gives a beautifully clear account of information theory [3] , and several mathematical models investigating the origin of the magical number seven have been reported [4] [5] . Whether the capacity of short-term memory is 4 - 5 or 7 ± 2 items, it is clearly limited. However, memory capacity can be increased through a process called chunking [1]. For example, in Japan, 11-digit cellular phone numbers and 10-digit toll free numbers are chunked into groups of three or four digits: 090-XXXX- XXXX and 0120-XXX-XXX, respectively. Phone numbers in many other countries are similarly chunked. It is unclear how many items per group provide the most efficient chunking, and the current study used probability theory to investigate this.

In probability theory, there is a problem entitled "the tourist with a short memory" [6]. For example, if a tourist wants to visit four capitals A, B, C, and D, he travels first to one capital chosen at random. If he visits A, the next time, he should choose among B, C, and D with the same probability. However, in this problem, the tourist quickly forgets that he has already visited A. Therefore, if he visits B second, the next time, he would choose among A, C, and D with the same probability. The problem is to find the expected number, E (N), of trips required until the tourist has visited all four capitals. To address this question, the problem is transformed into a problem of short-term memory based on some hypotheses and assumptions described below. In the present paper these capitals correspond to items which are recalled. We study a case without chunking (Procedure 1), a case in which items are chunked in order into groups containing the same number of items (Procedure 2), and a case in which items are chunked in order into groups containing the different number of items (Procedure 3). The novelty of this study is that the most effective chunking involves groups of three or four items, such as in phone numbers, and that 23 trips may be the critical number, beyond which some items will be forgotten. A 16-digit credit card number exceeds the capacity of short-term memory, even when chunked into groups of four digits, such as XXXX-XXXX-XXXX-XXXX. Based on these data, 16-digit credit card numbers should be sufficient for security purposes.

MODEL

When a subject responds to an event involving several stimuli, those stimuli must be processed in such a way to distinguish among them while still associating them with the entire set of items. According to Miller's and Cowan's hypotheses (7 ± 2 or 4 - 5 items, respectively) [1] [2], the capacity of short-term memory is between 4 - 9 items. Stimuli are often processed in order of dominance. The simplest way to order n items is to compare two items, retain the more dominant of the pair, then compare that with another item, again retaining the dominant one, and repeating this process until the entire collection has been ordered [5]. Although this process may be considered fundamental, it is assumed for simplicity that input items are one-dimensional categories, for example, strings of digits, letters, or words. The following assumptions were made:

Assumption 1: Input items (or stimuli) are assumed to be labeled as A_1, A_2, ..., and A_n in order.

Assumption 2: Items are remembered equally with no one item being more dominant. The probability to recall any A_j except A_i next after A_i is recalled is equal.

Assumption 3: The subject can only recall n items in order after he recalls every item at least once.

Applying these assumptions to the problem of "the tourist with a short memory", the problem is to find the expected number, E (N), of trips required until the tourist has visited all capitals. The process that any A_j except A_i is recalled after A_i is represented as a way: $W (A_i \rightarrow A_j)$. This can be calculated without chunking (Procedure 1) and with chunking into same-sized or different-sized groups (Procedures 2 and 3) as follows:

Procedure 1: To find the expected number, $E(N)$, of ways required until all A_i's are recalled (Figure 1(a)).

Procedure 2: Items are chunked in order into groups, which have the same number of items (Figure 1(b)). For example, (A_1, A_2, A_3), (A_4, A_5, A_6), (A_7, A_8, A_9), ..., and (A_{n-2}, A_{n-1}, A_n). Groups are denoted in order as $B_1, B_2, B_3, ...$ (Figure 1(b)). There is equal probability to recall any B_j except B_i immediately after B_i is recalled. When any B_j is recalled for the first time, all items in B_j are recalled at least once, which assumes that the relationship among the items in B_j has already been confirmed. Hence, all visits within B_j are remembered from the second visit of B_j onwards. When all B_i's are recalled, all A_i's are also recalled, confirming the relationship among all A_i's.

Procedure 3: Items are chunked into groups with different numbers of items. For example, in Japan, 11-digit cellular phone numbers and 10-digit toll free numbers are displayed as 090-XXXX-XXXX and 0120-XXX- XXX, respectively. The 11-digit phone number is chunked into three groups, B_1, B_2, B_3, one of which consists of three digits, $B_1 = (A_1,$

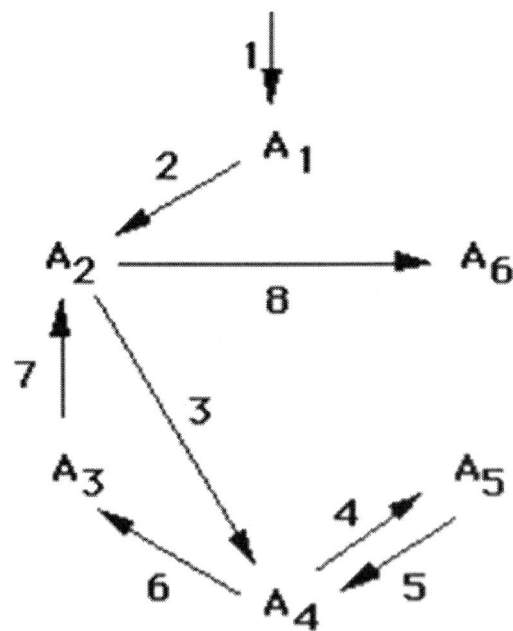

(a)

Probability Theory Predicts that Chunking into groups of three

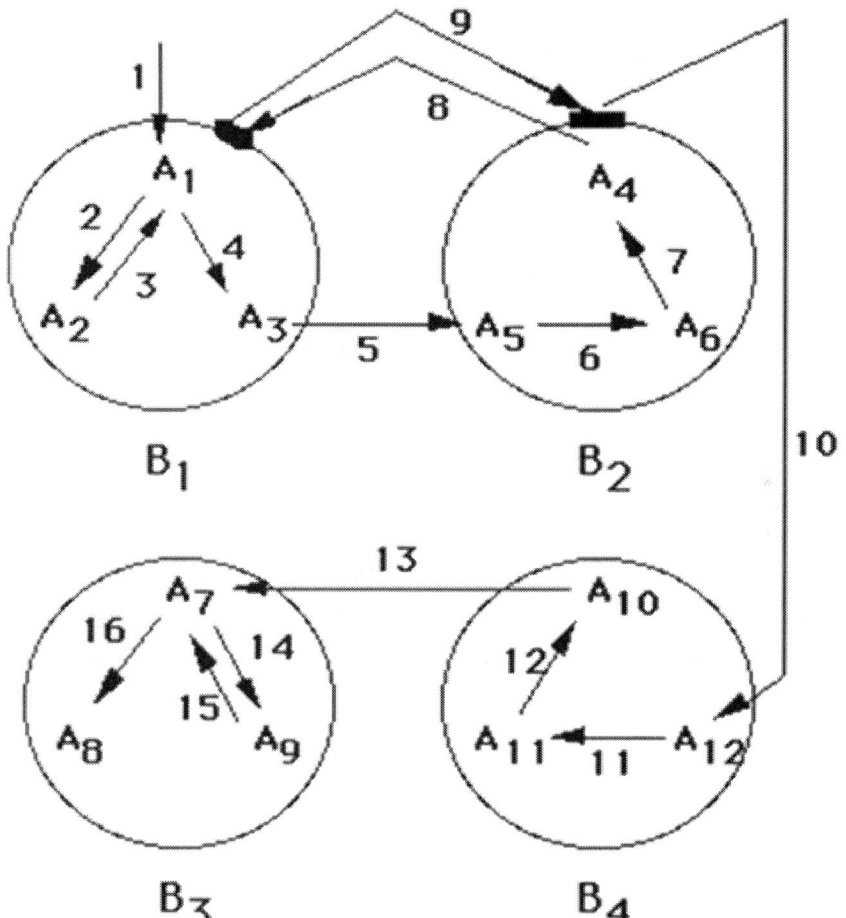

(b)

Figure 1: Ways, W(A$_i$→A$_j$), (labeled by turns) required until all A$_i$'s are recalled without any chunking of items (a) and with chunking of items into, for example, four groups (B$_1$, B$_2$, B$_3$, B$_4$) (b).

A$_2$, A$_3$), and two of which consist of four digits, B$_2$ = (A$_4$, A$_5$, A$_6$, A$_7$), B$_3$ = (A$_8$, A$_9$, A$_{10}$, A$_{11}$). Similarly, the 10-digit toll free number is chunked into three groups, B$_1$, B$_2$, B$_3$, one of which consists of four digits, B$_1$ = (A$_1$, A$_2$, A$_3$, A$_4$), and two of which consist of three digits, B$_2$ = (A$_5$, A$_6$, A$_7$), B$_3$ = (A$_8$, A$_9$, A$_{10}$).

RESULTS OF CALCULATION

Procedure 1

When the number of all items is n, the expected number, E(N), of ways, $W(A_i{\rightarrow}A_j)$, required until all A_i's are recalled can be calculated.

In the case of n = 3, a subject wants to recall three items, A_1, A_2, A_3. Set N as follows:

$$N = Y_0 + Y_1 + Y_2,$$

where Y_m is the number of ways required for recalling one more item when m items have already been recalled. Therefore, Y_m's are independent stochastic variables. Y_0 and Y_1 are always 1. $Y_0 = 1$ indicates the first way of recalling one of the items. For example, it corresponds to the first way of Figure 1(a). In case of Y_2, one item has yet not been recalled, but it is recalled the kth time with a geometric probability of

$p(1-p)^{k-1}; p = 1/2$ for k = 1, 2, ... The expected distribution is 1/p. Therefore, $E(Y_2) = 2$. Since Y_m's are mutually independent random variables, $E(N) = E(Y_0) + E(Y_1) + E(Y_2) = 1+1+2=4$.

This equation is transformed into $E(N)=1+2.\left(\dfrac{1}{2}+1\right)$.

When $Y_2 = y_2$, the probability of N; $P(N:Y_2 = y_2)$, is expressed as $P(N:Y_2=y_2)$ $= P(Y_0=1).P(Y_1=1).P(Y_2=y_2).P(Y_0=1)$ and $P(Y_1 = 1)$ are always 1.

$$P\left(Y_2 = y_2\right) = \left(\frac{1}{2}\right)^{y_2-1} \cdot \frac{1}{2} = \frac{1}{2^{y_2}}.$$

Therefore,

$$P\left(N:Y_2 = y_2\right) = \frac{1}{2^{y_2}}.$$

As $N=1+1+y_2$

$$P(N : Y_2 = y_2) = \frac{1}{2^{N-2}}.$$

Hence,

$$P(N) = \left(\frac{1}{2}\right)^{N-2}.$$

In the case of n = 4, a subject wants to recall four items, A_1, A_2, A_3, A_4. Set N as follows:

$$N = Y_0 + Y_1 + Y_2 + Y_3,$$

where Y_i is the number of ways required for recalling one more item when i items have already been recalled. Therefore, Y_i's are mutually independent random variables. Y_0 and Y_1 are always 1. In the case of Y_2, two items have not yet been recalled, so one of these two items is

recalled the kth time with a geometric probability of $p(1-p)^{k-1}$; $p = 2/3$ for k = 1, 2, ... Similarly, Y_3 has a geometric probability function with p = 1/3. The expected distribution of a geometric probability function is 1/p. Therefore, $E(Y_2) = 3/2$ and $E(Y_3) = 3$. Since Y_i's are mutually independent random variables,

$$E(N) = E(Y_0) + E(Y_1) + E(Y_2) + E(Y_3) = 1 + 1 + \frac{3}{2} + 3 = \frac{13}{2}.$$

This equation is transformed into $E(N) = 1 + 3\left(\frac{1}{3} + \frac{1}{2} + 1\right)$.

Therefore, the expression for a general number, n, of items is:

$$E(N) = 1 + (n-1)\left(\frac{1}{n-1} + \frac{1}{n-2} + \cdots + \frac{1}{2} + 1\right).$$

This can be easily proven.

When $Y_2 = y_2$ and $Y_3 = y_3$, the probability of N; $P(N: Y_2 = y_2, Y_3 = y_3)$, is expressed as $P(N:Y_2=y_2,Y_3=y_3) = P(Y_0=1).P(Y_1=1).P(Y_2=y_2).P(Y_3=y_3)$ $P(Y_0 = 1)$ and $P(Y_1 = 1)$ are always 1.

$$P(Y_2 = y_2) = \left(\frac{1}{3}\right)^{y_2-1} \cdot \frac{2}{3} = \frac{2}{3^{y_2}}.$$

$$P(Y_3 = y_3) = \left(\frac{2}{3}\right)^{y_3-1} \cdot \frac{1}{3} = \frac{2^{y_3-1}}{3^{y_3}}.$$

Therefore,

$$P(N : Y_2 = y_2, Y_3 = y_3) = \frac{2^{y_3}}{3^{y_2+y_3}}.$$

As $N=1+1+y_2+y_3$,

$$P(N : Y_2 = y_2, Y_3 = y_3) = \frac{2^{y_3}}{3^{N-2}}; 1 \le y_3 \le N-3.$$

Hence,

$$P(N) = \sum_{y_3=1}^{N-3} \frac{2^{y_3}}{3^{N-2}} = \left(\frac{1}{3}\right)^{N-2} \left(2^{N-2} - 2\right).$$

In the case of $n = 5$, $N=1+1+y_2+y_3+y_4$,

$$P(N) = \sum_{y_3=1}^{N-3} \frac{2^{y_3}}{3^{N-2}} = \left(\frac{1}{3}\right)^{N-2} \left(2^{N-2} - 2\right).$$

For a general number, n (≥3), of items,

$$P(N) = \left(\frac{1}{n-1}\right)^{N-2}$$

$$\left[\binom{n-2}{n-2} \cdot (n-2)^{N-2} + (-1)^1 \cdot \binom{n-2}{n-3} \cdot (n-3)^{N-2} + (-1)^2 \cdot \binom{n-2}{n-4} \cdot (n-4)^{N-2} + \cdots + (-1)^{n-3} \cdot \binom{n-2}{1} \cdot 1^{N-2}\right]$$

The equation is proved. Specifically, in the case of $n = 2$, $E(N) = 2$ with a probability of 1; in the case of $n = 3$, $E(N) = 4$, and the cumulative probability that N is smaller than or equal to $E(N)$, $P(N \leq E(N))$, is 0.75; in the case of $n = 4$, $E(N) = 13/2$ and $P(N \leq E(N)) = 0.7407$.

In the case of $n = 5$, which corresponds to one of Miller's magical numbers, $5(=7-2)$, $E(N) = 28/3$? 10 and in case of $n = 9$, which corresponds to the other of Miller's magical numbers, $9(=7+2)$, $E(N) = 796/35$? 23. In the case of $n = 10$, $E(N) = 7409/280$? 27. Clearly, as n increases, $E(N)$ increases exponentially (Figure 2(a)).Hence, the greater the number, n, of items, the greater the difficulty to recall all items. Although the cumulative probability of $P(N \leq E(N))$ decreases steadily, it is larger than 0.5 until $n = 40$ (Figure 2(b)).

Procedure 2

Items are chunked in order into groups with all groups containing the same number of items. The number of all items is denoted as n, and the number of items in each group is denoted as m. For an example of $m = 3$, the groups are (A_1, A_2, A_3), (A_4, A_5, A_6), (A_7, A_8, A_9), ... (A_{n-2}, A_{n-1}, A_n). These groups are denoted in order as B_1, B_2, B_3, ... (Figure 1(b)). Similar to Procedure 1, there is equal probability to recall any B_j except B_i immediately after B_i. When any B_j is recalled for the first time, all items in B_j are recalled at least once, so it is assumed that the relationship among the items in B_j has already been confirmed. Hence, all visits within B_j are saved from the second visit of B_j onwards. When all B_i's are recalled, it means that all A_i's are recalled, confirming the relationship among all A_i's.

The number of B_i's is $\frac{n}{m}$, which is replaced by the nearest integer above

$\frac{n}{m}$, $\left[\frac{n}{m}\right]$, if $\frac{n}{m}$ is not an integer.

The expected number, $E(N_{n, m})$, of ways required until all A_i's are recalled can be calculated. For the example of $n = 12$ and $m = 3$, a subject wants to recall 12 items, A_1, A_2, A_3, ... ,A_{12}. Then, $B_1 = (A_1, A_2, A_3)$, $B_2 = (A_4, A_5, A_6)$, $B_3 = (A_7, A_8, A_9)$, and $B_4 = (A_{10}, A_{11}, A_{12})$.

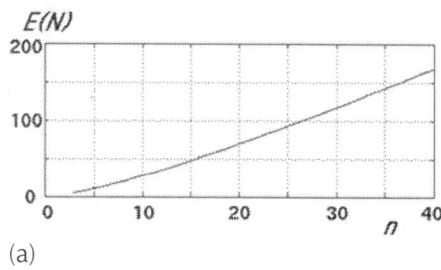

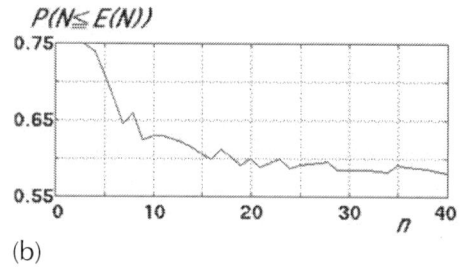

Figure 2: (a) The expectation of the number, E(N), of ways, W($A_i \rightarrow A_j$), required until all A_i's are recalled; (b) The cumulative probability that N is smaller than or equal to E(N), P(N $\leq$ E(N)). n represents the number of items.

Set $N_{12,3}$ as follows:

$$N = Z_0 + Z_1 + Z_2 + Z_3 + (Y_1 + Y_2) \times 4,$$

where Z_i is the number of ways required for recalling one more group when i groups have been recalled, and Y_j is the number of ways required for recalling one more item of any group when j items of this group have been recalled. Therefore, Z_i's and Y_j's are mutually independent random variables. Z_0, Z_1, and Y_1 are always 1. Specifically, Z_0 = 1 indicates the first way going to one of the groups. For example, it corresponds to the first way of Figure 1(b). Hence,

$$E(N_{12,3}) = E(Z_0) + E(Z_1) + E(Z_2) + E(Z_3) + \left[E(Y_1) + E(Y_2)\right] \times 4.$$

Using the case of four items in Procedure 1, we can regard the four groups in Procedure 2 as four items,

$$E(Z_0) + E(Z_1) + E(Z_2) + E(Z_3) = \frac{13}{2}.$$

Using the case of three items from Procedure 1,

$$E(Y_1) + E(Y_2) = 1 + 2 = 3.$$

Hence,

$$E(N_{12,3}) = \frac{37}{2}.$$

As another practical example, the expected number of ways required to recall 16 digits, $E(N_{16,4})$, corresponding to a credit card account number, XXXX-XXXX-XXXX-XXXX, can be calculated.

$$E(N_{16,4}) = E(Z_0) + E(Z_1) + E(Z_2) + E(Z_3) + \left[E(Y_1) + E(Y_2) + E(Y_3) \right] \times 4.$$

Using the case of four items in Procedure 1 and regarding the four groups as four items,

$$E(Z_0) + E(Z_1) + E(Z_2) + E(Z_3) = \frac{13}{2}.$$

Using the case of four items in Procedure 1,

$$E(Y_1) + E(Y_2) + E(Y_3) = \frac{11}{2}.$$

Hence,

$$E(N_{16,4}) = \frac{57}{2}.$$

Generally,

$$E(N_{n.m}) = \sum_{i=0}^{m} E(Z_i) + \left[\frac{n}{m} \right] \times \sum_{j=1}^{\left[\frac{n}{m} \right]} E(Z_i).$$

Then, if $\frac{n}{m}$ is an integer, $\left[\frac{n}{m} \right] = \frac{n}{m}$, otherwise $\left[\frac{n}{m} \right]$ stands for the nearest integer above $\frac{n}{m}$. $E(N_{n,m})$ can only be calculated precisely when n is a multiple of m. However, even if n is not a multiple of m, $E(N_{n,m})$ is calculated to observe the relationship between m and $E(N_{n,m})$. This calculation will be justified when n is larger than m, for example $n \geq 20$ and $1 \leq m \leq 10$. When $10 \leq n \leq 20$., $E(N_{n,m})$ is calculated only when n is a multiple of m. $E(N_{n,m})$ is calculated for n = 10, 11, ..., 100, and m = 1, 2, ..., 10. Figure 3 shows the results for n = 20, 30, 40, ..., 100 and m =

1, 2, ..., 10. When m = 3 and 4, $E(N_{n,m})$ is the smallest and the second smallest for any n(10≤n≤100). When m = 2 or 5, $E(N_{n,m})$ is the third smallest. It is interesting to note that the case of m = 1 corresponds to any case without chunking from Procedure 1.

Procedure 3

The expected number $E(N_{n,*})$ of ways required until all A_i's are recalled can be calculated in the same manner as Procedure 2 for special cases of items chunked into groups of different lengths. When lengths of chunked groups, m = 2, 3, or 4, $E(N_{n,m})$ is the smallest. All integers are expressed by a sum of 2's, 3's, and 4's.For example, 17=2+3+4×3. Hence, items of any length can be chunked into groups, the lengths of which are 2, 3, or 4.

The 11-digit phone number 090-XXXX-XXXX is chunked into three groups, B_1, B_2, B_3, one of which consists of three digits, $B_1 = (A_1, A_2, A_3)$, and two of which consist of four digits, $B_2 = (A_4, A_5, A_6, A_7)$, $B_3 = (A_8, A_9, A_{10}, A_{11})$.

$$E(N) = E(Z_0) + E(Z_1) + E(Z_2) + \left[E(Y_1) + E(Y_2) \right] + \left[E(Y_1) + E(Y_2) + E(Y_3) \right] \times 2.$$

Hence,

$$E(N) = 4 + 3 + \frac{11}{2} \times 2 = 18.$$

The 10-digit phone number 0120-XXX-XXX is chunked into three groups, B_1, B_2, B_3, one of which consists of four digits, $B_1 = (A_1, A_2, A_3, A_4)$, and two of which consist of three digits, $B_2 = (A_5, A_6, A_7)$, $B_3 = (A_8, A_9, A_{10})$.

$$E(N) = E(Z_0) + E(Z_1) + E(Z_2) + \left[E(Y_1) + E(Y_2) + E(Y_3) \right] + \left[E(Y_1) + E(Y_2) \right] \times 2.$$

Hence,

$$E(N) = 4 + \frac{11}{2} + 3 \times 2 = \frac{31}{2}.$$

Probability Theory Predicts that Chunking into groups of three

A 10-digit phone number of 03-XXXX-XXXX (for example, in Tokyo) is chunked into three groups, B_1, B_2, B_3, one of which consists of two digits, $B_1 = (A_1, A_2)$, and two of which consist of four digits, $B_2 = (A_3, A_4, A_5, A_6)$, $B_3 = (A_7, A_8, A_9, A_{10})$.

$$E(N) = E(Z_0) + E(Z_1) + E(Z_2) + E(Y_1) + \left[E(Y_1) + E(Y_2) + E(Y_3) \right] \times 2.$$

Hence,

$$E(N) = 4 + 1 + \frac{11}{2} \times 2 = 16.$$

DISCUSSION

Findings Obtained from the Mathematical Model
Without Chunking

As the number of the items, n, increases, the expected number, E(N), of ways required until all items are recalled increases exponentially. The cumulative probability that N is smaller than or equal to E(N), P(N ≤ E(N)), decreases steadily. Hence, the greater the number, n, of items,

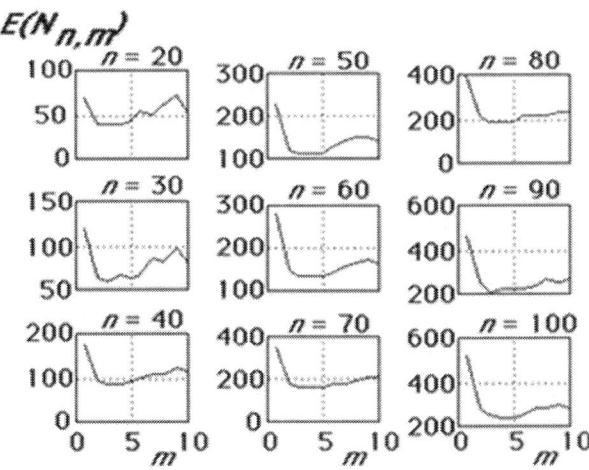

Figure 3: The expected number, $E(N_n, m)$, of ways required until all A_i's are recalled. n represents the number of items. m represents the number of chunked groups.

the greater the difficulty to recall all items. In the case of five items, which corresponds to one of Miller's magical numbers ($7 - 2 = 5$), $E(N)$ = 28/3 ? 10, and in the case of nine items, which corresponds to the other of Miller's magical numbers ($7 + 2 = 9$), $E(N) = 796/35$? 23. In the case of $n = 10$, $E(N) = 7409/280$? 27.

With Chunking

$E(N_{n,m})$ is the expected number of ways required until all items are recalled. Hence, a smaller value for $E(N_{n,m})$ indicates more efficient recall. For example, the expected number of ways required until 12 items chunked into three groups are recalled, $E(N_{12,3})$, is 37/2 ? 19. In the case of a 16-digit credit card number, XXXX-XXXX- XXXX-XXXX, $E(N_{16,4}) = 57/2$? 29. From the results for $n = 10, 11, \ldots, 100$, and $m = 1, 2, \ldots, 10$, $E(N_{n,m})$ is the smallest for any n ($10 \leq n \leq 100$), when $m = 3$ or 4. Hence, when $m = 3$ or 4, all items can be recalled most quickly.

Special cases of Items Chunked into Groups of Different Lengths

The expected number of ways required to recall all 11 digits (e.g., in the phone number 090-XXXX-XXXX), $E(N_{11,*})$, is 18. For a 10-digit phone number in the format 0120-XXX-XXX, $E(N_{10,*}) = 31/2$? 16. For a 10-digit phone number in the format 03-XXXX-XXXX, $E(N_{10,*}) = 16$.

Interpretation of the Findings
Without Chunking

Short-term memory lasts from several seconds to several minutes. Based on the current data, we conclude that an individual can follow the 23 ways required to recall nine items within several minutes, but it takes longer to follow the 27 ways required to recall 10 items, so some one of the items are forgotten. These results suggest that 23 ways may be the critical number, beyond which some items will be forgotten.

With Chunking

A smaller number of $E(N_{n,m})$ indicates more efficient recall. From the results for $n = 10, 11, \ldots, 100$, and $m = 1, 2, \ldots, 10$, $E(N_{n,m})$ is the smallest for any n, $(10 \leq n \leq 100)$ when m = 3 or 4. Each group has 3 or 4 items (m = 3 or 4) without chunking. From Procedure 1, $P(N \leq E(N))$ is 0.75 in the case of three items, and $P(N \leq E(N))$ is 0.7407 in the case of four items. $P(N \leq E(N))$ decreases steadily with more items. Hence, when m = 3 or 4, all items of each group can be recalled most quickly and with the greatest confidence. $E(N_{12,3})$ = 37/2 ? 19 is less than 23, the critical number for recall. Hence, chunking will be effective: B_1 = (A_1, A_2, A_3), B_2 = (A_4, A_5, A_6), B_3 = (A_7, A_8, A_9), B_4 = (A_{10}, A_{11}, A_{12}). However, for 16 digits, such as in a credit card number, XXXX-XXXX- XXXX-XXXX, $E(N_{16,4})$ = 57/2 ? 29, which is larger than the critical number for recall. Thus chunking will not benefit short-term memory recall of a 16-digit credit card number. Based on these findings, a 16-digit credit card number of XXXX-XXXX-XXXX-XXXX should have greater security than a 12-digit number of XXX-XXX- XXX-XXX.

Special cases of Items Chunked into Groups of Different Sizes

The expected numbers, E(N), of ways for 090-XXXX-XXXX, 0120-XXX-XXX, and 03-XXXX-XXXX, are less than 23, the critical number for recall. Hence, chunking into groups of two to four items is truly effective for recalling 11 or 10-digit phone numbers.

Study Limitations

The current findings were obtained using a model based on certain assumptions. The validity of these assumptions should be investigated in the future.

CONCLUSIONS

We use probability theory to predict that the most effective chunking involves groups of three or four items, such as in phone numbers, and conclude that an individual can follow the 23 ways required to recall nine items within several minutes, but it takes longer to follow the 27 ways required to recall 10 items, so some of the items are forgotten. These results suggest that 23 ways may be the critical number, beyond which some items will be forgotten. A 16-digit credit card number exceeds the capacity of short-term memory, even when chunked into groups of four digits, such as XXXX-XXXX-XXXX-XXXX. Based on these data, 16-digit credit card numbers should be sufficient for security purposes.

APPENDIX

The equation,

$$P(N) = \left(\frac{1}{n-1}\right)^{N-2}$$

$$\left[\binom{n-2}{n-2}\cdot(n-2)^{N-2} + (-1)^1\cdot\binom{n-2}{n-3}\cdot(n-3)^{N-2} + (-1)^2\cdot\binom{n-2}{n-4}\cdot(n-4)^{N-2} + \cdots + (-1)^{n-3}\cdot\binom{n-2}{1}\cdot 1^{N-2}\right],$$

for a general number, n ($\geq$3), of items, represents the probability, P(N), that all A_i's are not visited until the N-th way W ($A_j \to A_i$). Then A_i is visited lastly and only once. This equation is proved below.

Proof: Let see Figure 1(a). Then, A_1 is visited first, A_2 is visited second, and thereafter these may be visited several times. It is assumed that the first visit is A_1 and the second visit is A_2 without loss of generality. It is assumed that the last visit is A_i (i = 3, 4, …,n). When the present visit is A_j (j $\neq$ i), the probability that A_i is visited is $\frac{1}{n-1}$ and A_i is one of A_3, A_4, …, A_n except A_j (j $\neq$ i). The probability that the items except A_i are visited totally

N ? 3 times is $\binom{n-3}{n-3}\left(\frac{n-2}{n-1}\right)^{N-3}$ Hence, the probability C(0) that the items except A_i are visited totally N ? 3 times and A_i is visited lastly is

$$C(0) = \frac{1}{n-1}\times(n-2)\times\binom{n-3}{n-3}\left(\frac{n-2}{n-1}\right)^{N-3} = \binom{n-2}{n-2}\times(n-2)\times\frac{1}{n-1}\left(\frac{n-2}{n-1}\right)^{N-3} = \binom{n-2}{n-2}\left(\frac{n-2}{n-1}\right)^{N-2}$$

However, some events that at least k (1 $\leq$ k $\leq$ n ? 3) items except A_1, A_2, and A_i are not visited should be excluded. This probability C(k) is

Probability Theory Predicts that Chunking into groups of three

$$C(k) = \frac{1}{n-1} \times (n-2) \times \left(\frac{n-3}{n-3-k}\right)\left(\frac{n-2-k}{n-1}\right)^{N-3} = \left(\frac{n-3}{n-3-k}\right) \times (n-2) \times \frac{1}{n-1}\left(\frac{n-2-k}{n-1}\right)^{N-3}$$

$$= \left(\frac{n-2}{n-2-k}\right) \times (n-2-k) \times \frac{1}{n-1}\left(\frac{n-2-k}{n-1}\right)^{N-3} = \left(\frac{n-2}{n-2-k}\right)\left(\frac{n-2-k}{n-1}\right)^{N-2}.$$

Moreover, some events that at least m ($\leq$ n ? 3 ? k) items except A_1, A_2, A_i, and those excluded k items are not visited should be excluded. This probability C(k, m) is

$$C(k,m) = \frac{1}{n-1} \times (n-2) \times \left(\frac{n-3}{n-3-k}\right)\left(\frac{n-3-k}{n-3-k-m}\right)\left(\frac{n-2-k-m}{n-1}\right)^{N-3}$$

$$= \left(\frac{n-3}{n-3-k-m}\right) \times \left(\frac{k+m}{m}\right) \times \frac{n-2}{n-1}\left(\frac{n-2-k-m}{n-1}\right)^{N-3}.$$

D(0) is defined as C(0). D(k), (1 $\leq$ k $\leq$ n ? 3), is defined as D(k ? 1) + (?1) ^{k} C(k). D(p, p), (0 $\leq$ p $\leq$ n ? 3), is defined as the probability that at least p items except A_1, A_2, and A_i are not visited within D(p).

1) Since D(1) is also defined as D(0) ? C(1), D(1,1) = 0.

2) D(2) is also defined as D(0) ? C(1) + C(2).

$$D(2,2) = D(0,2) - C(1,1) + C(2,0) = \left(\frac{n-3}{n-5}\right) \times \left(\frac{2}{2}\right) \times \frac{n-2}{n-1}\left(\frac{n-4}{n-1}\right)^{N-3} - \left(\frac{n-3}{n-5}\right) \times \left(\frac{2}{1}\right) \times \frac{n-2}{n-1}\left(\frac{n-4}{n-1}\right)^{N-3}$$

$$+ \left(\frac{n-3}{n-5}\right) \times \left(\frac{2}{0}\right) \times \frac{n-2}{n-1}\left(\frac{n-4}{n-1}\right)^{N-3} = \left(\frac{n-3}{n-5}\right) \times \left[\left(\frac{2}{2}\right) - \left(\frac{2}{1}\right) + \left(\frac{2}{0}\right)\right] \times \frac{n-2}{n-1}\left(\frac{n-4}{n-1}\right)^{N-3}$$

$$= \left(\frac{n-3}{n-5}\right) \times (1-1)^2 \times \frac{n-2}{n-1}\left(\frac{n-4}{n-1}\right)^{N-3} = 0$$

3) D(k) = D(k?1)+ (?1)^{k}C(k): 1 $\leq$ k $\leq$ n?3.

$$D(k,k) = D(0,k) - C(1,k-1) + C(2,k-2) + \cdots + (-1)^k C(k,0)$$

$$= \sum_{i=0}^{k} (-1)^i \binom{n-3}{n-2-k} \times \binom{k}{k-i} \times \frac{n-2}{n-1} \left(\frac{n-2-k}{n-1}\right)^{N-3} = \binom{n-3}{n-2-k} \times \frac{n-2}{n-1} \left(\frac{n-2-k}{n-1}\right)^{N-3} \sum_{i=0}^{k} (-1)^i \binom{k}{k-i}$$

$$= \binom{n-3}{n-2-k} \times \frac{n-2}{n-1} \left(\frac{n-2-k}{n-1}\right)^{N-3} \times (1-1)^k = 0$$

Hence, D(k, k) = 0 (1 ≤ k ≤ n ? 3). Hence, the probability that at least q, (1 ≤ q ≤ n ? 3), items except A_1, A_2, and A_i are not visited within D(k) is equal to 0. In other words, D(n ? 3) represents the probability that all items except A_i are visited totally N ? 1 times and A_i is visited lastly.

$$D(n-3) = D(0) - C(1) + C(2) + \cdots + (-1)^{n-3} C(n-3)$$

$$= C(0) - C(1) + C(2) + \cdots + (-1)^{n-3} C(n-3) = P(N).$$

The equation has been proved.

REFERENCES

1. Miller, G.A. (1956) The Magical Number Seven plus or minus Two: Some Limits on Our Capacity for Processing Information. Psychological Review, 63, 81-97. http://dx.doi.org/10.1037/h0043158.
2. Cowan, N. (2001) The Magical Number 4 in Short-Term Memory: A Reconsideration of Mental Storage Capacity. Behavioral and Brain Sciences, 24, 87-114.http://dx.doi.org/10.1017/S0140525X01003922.
3. Baddely, A. (1994) The Magical Number Seven: Still Magic after All These Years. Psychological Review, 101, 353- 356. http://dx.doi.org/10.1037/0033-295X.101.2.353.
4. Nicolis, J.S. and Tsuda, I. (1985) Chaotic Dynamics of Information Processing: The "Magic Number Seven Plus-Minus Two" Revisited. Bulletin Mathematical Biology, 47, 343-365.
5. Satty, T.L. and Ozdemir, M.S. (2003) Why the Magic Number Seven plus or minus Two. Mathematical and Computer Modelling, 38, 233-244. http://dx.doi.org/10.1016/S0895-7177(03)90083-5.
6. Blom, G., Holst, L. and Sandell, D. (1991) Problems and Snapshots from the World of Probability. Springer-Verlag, New York.

CITATION

1. Osaka, M. (2014) Probability Theory Predicts That Chunking into Groups of Three or Four Items Increases the Short-Term Memory Capacity.Applied. Mathematics, 5, 1474-1484 doi: 10.4236/am.2014.510140.

Applications to Cryptography of Twisting Commutative Algebraic Groups

A. Silverberg
Mathematics Department, University of
California, Irvine, CA 92697-3875, USA

ABSTRACT

We give an overview on twisting commutative algebraic groups and applications to discrete log-based cryptography. We explain how discrete log-based cryptography over extension fields can be reduced to cryptography in primitive subgroups. Primitive subgroups in turn are part of a general theory of tensor products of commutative algebraic groups and Galois modules (or twists of commutative algebraic groups), and this underlying mathematical theory can be used to shed light on discrete log-based cryptosystems. We give a number of concrete examples, to illustrate the definitions and results in an explicit way.

INTRODUCTION

In this paper we give a survey on twisting commutative algebraic groups and applications to discrete log-based cryptography. One of our goals will be to explain part of the paper [23] at a more down-to-earth, less technical level, and explain some of its connections to cryptography, in order to make these ideas accessible to a wider audience of mathematicians and cryptographers. We hope that this more general setting will lead to a better understanding of known cryptosystems and their underlying mathematics, and possibly lead to new ideas. We give an overview; see the cited papers for details. In particular, see [23] for most of the results stated in this paper.

A number of cryptosystems, including the Lucas-based [24], [36], [37], [41] and [42], Gong–Harn [13], XTR [2] and [20], and $\mathbb{T}_2$ and CEILIDH [27] cryptosystems, and the abelian variety or elliptic curve systems in [26], can be viewed as being based on the idea that when one does discrete log-based cryptography (for either a multiplicative group of a field or an elliptic curve group) over field extensions, one can improve bandwidth efficiency by restricting to a suitable "primitive" subgroup.

Let $\mathbb{F}_q$ denote the finite field with q elements. Discrete log-based cryptography is generally performed using the $\mathbb{F}_q$-points of the multiplicative group $\mathbb{G}_m$, or the $\mathbb{F}_q$-points of an abelian variety A over $\mathbb{F}_q$ (usually an elliptic curve or the Jacobian variety of a hyperelliptic curve). Cryptography over an extension field $\mathbb{F}_{q^n}$ with n>1 uses the groups

$$\mathbb{G}_m(\mathbb{F}_{q^n}) = \mathbb{F}_{q^n}^{\times} = \mathbb{F}_{q^n} - \{0\}$$

or $A(\mathbb{F}_{q^n})$, where A is an abelian variety over $\mathbb{F}_{q^n}$. The group of $\mathbb{F}_{q^n}$-points of $\mathbb{G}_m$ or A is isomorphic to the group of $\mathbb{F}_q$-points of the Weil restriction of scalars (see Definition 2.2) of $\mathbb{G}_m$ or A from $\mathbb{F}_{q^n}$ down to $\mathbb{F}_q$. So the Weil restriction of scalars arises naturally when doing discrete log-based cryptography over extension fields (see [7], [25], [5], [6], [12] and [9]).

Letting $V = \mathbb{G}_m$ or A, where A is now an abelian variety over the ground field $\mathbb{F}_q$, then $V(\mathbb{F}_{q^n})$ has a natural decomposition (via a homomorphism with "controlled" kernel and cokernel) into a direct product of groups $V_d(\mathbb{F}_q)$, over all divisors d of n (see Proposition 5.3). Thus, when doing finite field or abelian variety cryptography over $\mathbb{F}_{q^n}$, it suffices to consider the subgroups $V_d(\mathbb{F}_q)$ for all divisors d of n. Here, V_d is an algebraic torus over $\mathbb{F}_q$ if $V = \mathbb{G}_m$ and is an abelian variety over $\mathbb{F}_q$ if V is an abelian variety. Over $\mathbb{F}_{q^d}$, V_d is isomorphic to $V^{\phi(d)}$, where ϕ is the Euler ϕ-function, so dim $(V_d) = \phi$ (d) dim (V). When $V = \mathbb{G}_m$, then the group $V_d(\mathbb{F}_q)$ is isomorphic to the subgroup of $\mathbb{F}_{q^d}$ order $\Phi_d(q)$, where $\Phi_d(x) \in \mathbb{Z}[x]$ is the cyclotomic polynomial whose complex roots are the primitive d^{th} roots of unity.

It is useful to view these two settings, multiplicative groups of finite fields and abelian varieties over finite fields, as part of a general framework, as in [23]. Instead of dealing separately with $\mathbb{G}_m$ and A, and only considering finite fields, we will consider the more general setting of commutative algebraic groups over arbitrary fields. This generality allows us to consider the two settings of interest in cryptography, namely $\mathbb{G}_m(\mathbb{F}_{q^n})$ and A ($\mathbb{F}_{q^n}$), simultaneously, and to view them in the same general framework. One point of this paper is to show how the general framework in [23] allows one to recover (known) results that were previously dealt with separately in the two cases. Readers who are uncomfortable with the abstract theory are advised to restrict to the multiplicative group and elliptic curves over finite fields.

Sections 2 and 5 below give an overview of the primitive subgroups and their properties. Section 3 includes general definitions of the twisted commutative algebraic group $\mathcal{I} \otimes_O V$ arising from a commutative algebraic group V and a suitable Galois module $\mathcal{I}$, as given in [23]. Special cases of the varieties $\mathcal{I} \otimes_O V$ include powers, restrictions of scalars, twists of elliptic curves or abelian varieties, and primitive subgroups, including the algebraic tori that arise in torus-based cryptography and the abelian varieties that arise in [26]. In Section 4 we discuss the decomposition of the group rings $\mathbb{Q}[G]$, for G a finite abelian group, that gives rise to the decomposition of the restriction of scalars into primitive subgroups, and work out a concrete example in Section 5.6. In Section 6 on open problems we encourage work on questions of efficiently representing elements of primitive subgroups, and cryptographic security.

Interspersed throughout the paper are a number of examples. In particular, we use the cases of quadratic twists of elliptic curves and algebraic tori associated to quadratic extensions to illustrate a number of definitions and properties. See [26], [27] and [29] for additional examples.

Primitive subgroups associated with abelian varieties were discussed in a cryptographic context in [8] and [26], and have also arisen in work on polarizations on abelian varieties [16], constructing abelian varieties over number fields with Shafarevich–Tate groups of nonsquare order [38], and bounding below the Selmer rank of abelian varieties over dihedral extensions of number fields [22]. See [4] and [10] for the setting of generalized Jacobians.

Related tensor product constructions were given in [15] (see Proposition 12.7 on p. 205), Section 2 of [31] (when V is an elliptic curve with complex multiplication by $\mathcal{O}$ and $\mathcal{I}$ is a projective $\mathcal{O}$-module with trivial Galois action), [21] (for abelian varieties) and Section 7 of [3] (when V is a group scheme with $\mathcal{O}$-action and $\mathcal{I}$ is a projective $\mathcal{O}$-module with trivial Galois action). In the case of abelian varieties, the restriction of scalars was decomposed into primitive pieces in [6] (see also [5] and [25]). We note that the relevant parts of [6] hold without change for arbitrary commutative algebraic groups.

As usual, $\mathbb{Z}$, $\mathbb{Q}$, and $\mathbb{C}$ denote the integers, rational numbers, and complex numbers. If R is a commutative ring and G is a finite group, let

$$R[G] := \left\{ \sum_{g \in G} a_g g : a_g \in R \right\}.$$

PRIMITIVE SUBGROUPS

In Definition 2.4 below we will give several equivalent definitions of primitive subgroups. We begin by defining algebraic groups and the Weil restriction of scalars.

Definition 2.1

An algebraic group or group variety is an algebraic variety V together with a "group operation" morphism G×G→G and an "inverse" morphism G→G with respect to which G is a group. A homomorphism of algebraic groups is a morphism of varieties that is also a group homomorphism.

Suppose in this section that V is a commutative algebraic group over a field k, and L is an abelian extension of k of finite degree n. We will view the group law on V multiplicatively. When restricting to examples, this is fine when V is the multiplicative group $\mathbb{G}_m$, but unfortunately is perhaps confusing when V is an elliptic curve, where one normally uses additive notation.

Restriction of Scalars

Definition 2.2

The restriction of scalars of V from L down to k, denoted $\mathrm{Res}_k^L(V)$ or $\mathrm{Res}_{L/k}(V)$ (we will use the former notation), is a commutative algebraic group over k along with a homomorphism

$$\eta_{L/k} : \mathrm{Res}_k^L(V) \to V$$

defined over L, with the universal property that for every variety X over k, the map

$$\mathrm{Hom}_k(X, \mathrm{Res}_k^L(V)) \xrightarrow{\sim} \mathrm{Hom}_L(X, V), \quad f \mapsto \eta_{L/k} \circ f \qquad (2.1)$$

is an isomorphism.

For every k-algebra A, it follows (by taking $X = \mathrm{Spec}(A)$) that $\eta L/k$ induces an isomorphism $\eta L/k : (\mathrm{Res}_k^L(V))(A) \xrightarrow{\sim} V(A \otimes_k L)$. In particular,

$$(\mathrm{Res}_k^L(V))(k) \cong V(L).$$

(2.2)

Further, if $k \subseteq M \subseteq L$ then

$$\mathrm{Res}_k^M(\mathrm{Res}_M^L(V)) = \mathrm{Res}_k^L(V).$$

(2.3)

(See Section 1.3 of [40] or Section 3.12 of [39] for background on the restriction of scalars.)

Remark 2.3: More concretely, if V is defined by a system of polynomial equations

$$f_i(x_1, \ldots, x_r) = 0, \quad 1 \le i \le s$$

with coefficients in k (or more generally, L), fix a basis $\{v_1, \ldots, v_n\}$ for L over k, and write $x_i = \sum_{j=1}^n y_{ij} v_j$ with variables y_{ij}. Substitute this into the equations $f_i(x_1, \ldots, x_r) = 0$, expand, and equate coefficients of the basis vectors $\{v_1, \ldots, v_n\}$, to obtain a system of polynomials in the variables $\{y_{ij}\}$ with coefficients in k. The n·dim (V)-dimensional variety over k defined by these equations is $\mathrm{Res}_k^L(V)$.

Next, we give two examples. We will use these examples throughout the paper, as a simple way to illustrate the mathematical definitions and results. See [26], [27] and [29] for other examples.

The Multiplicative Group and Quadratic Extensions

Suppose k is a field whose characteristic is not 2, and suppose $D \in k^\times$ is a nonsquare. Let $L = k(\sqrt{D})$, let $G = \mathrm{Gal}(L/k)$, and let σ be the generator of G. It is standard to view the multiplicative group $\mathbb{G}_m$ as the variety in $\mathbb{A}^2$ defined by the equation xy=1, which is the same as identifying $\mathbb{G}_m$ with the nonzero part of the x-line. Then $\mathrm{Res}_k^L(\mathbb{G}_m)$ is the variety R in $\mathbb{A}^3$ defined by the equation $(x_1^2 - Dx_2^2)y = 1$, which is an algebraic group with multiplication $(x_1, x_2, y) \cdot (w_1, w_2, z) = (x_1 w_1 + D x_2 w_2, x_1 w_2 + x_2 w_1, yz)$ and identity element $(x_1, x_2, y) = (1, 0, 1)$. That $\mathrm{Res}_k^L(\mathbb{G}_m) = R$ can be seen as follows. Define

$$\eta_{L/k} : R \to \mathbb{G}_m, \qquad (x_1, x_2, y) \mapsto (x_1 + x_2\sqrt{D}, (x_1 - x_2\sqrt{D})y)$$

(or more simply, $(x_1, x_2, y) \mapsto x_1 + x_2\sqrt{D}$). If X is a variety over k and $\psi \in \mathrm{Hom}_L(X, \mathbb{G}_m)$, define

$$\tilde{\psi} = \left(\frac{\psi + \psi^\sigma}{2}, \frac{\psi - \psi^\sigma}{2\sqrt{D}}, \frac{1}{\psi \psi^\sigma} \right) \in \mathrm{Hom}_k(X, R).$$

It is easy to check that $\eta_{L/k} \circ \tilde{\psi} = \psi$. If $f \in \mathrm{Hom}_k(X, R)$, it is easy to check that $\widetilde{(\eta_{L/k} \circ f)} = f$. It follows that the map $f \mapsto \eta_{L/k} \circ f$ of (2.1) is an isomorphism.

Elliptic Curves and Quadratic Extensions

Suppose $E : y^2 = f(x)$ is an elliptic curve over a field k whose characteristic is not 2, with deg $(f) = 3$, and suppose $D \in k^\times$ is a nonsquare. Let $L = k(\sqrt{D})$, let $G = \mathrm{Gal}(L/k)$, and let σ be the generator of G. Let $E^{(D)}$ be the elliptic curve $Dy^2 = f(x)$, the quadratic twist of E by D. Define

$$\phi : E \xrightarrow{\sim} E^{(D)}, \qquad (x, y) \mapsto (x, y/\sqrt{D}), \tag{2.4}$$

an isomorphism defined over L. Note that $\phi^\sigma = -\phi$. We claim that $\mathrm{Res}_k^L(E)$ is $(E \times E^{(D)})/T$, where

$$T = \{(P, \phi(P)) \in E \times E^{(D)} : 2P = O\} = \ker(f_0) \cap \ker(2)$$

with $f_0 : E \times E^{(D)} \to E$ the map that sends (P,Q) to $P - \phi^{-1}(Q)$. Here,

$$\eta_{L/k} : (E \times E^{(D)})/T \to E, \qquad (P, Q) \mapsto P + \phi^{-1}(Q).$$

To see that the universal property defining $\mathrm{Res}_k^L(E)$ holds for $(E \times E^{(D)})/T$ with this $\eta L/k$, suppose X is a variety over k and $\psi \in \mathrm{Hom}_L(X,E)$, note that multiplication by 2 induces an isomorphism $[2]: E/E[2] \tilde{\to} E$, let

$$\varphi = [2]^{-1} \circ \psi \in \mathrm{Hom}_L(X, E/E[2]), \text{ define}$$

$$\lambda : E/E[2] \to (E \times E^{(D)})/T, \qquad P \mapsto (P, \phi(P)) \bmod T,$$
$$\tilde{\psi} := \lambda \circ \varphi + (\lambda \circ \varphi)^\sigma \in \mathrm{Hom}_k(X, (E \times E^{(D)})/T),$$

and check that $\eta L/k \circ \lambda \circ \varphi = \psi$ and $\eta L/k \circ \lambda^\sigma \circ \varphi^\sigma = 0$, and thus

$$\eta_{L/k} \circ \tilde{\psi} = \psi.$$

We leave as an exercise to check that

$$\widetilde{(\eta_{L/k} \circ f)} = f$$

for every $f \in \mathrm{Hom}_k(X, (E \times E^{(D)})/T$. It follows that the map $f \mapsto \eta L/k \circ f$ of (2.1) is an isomorphism.

More concretely, one can write down a system of equations for $\mathrm{Res}_k^L(E)$ as follows. If E is $y^2 = x^3 + ax + b$, then substituting $x = x_1 + x_2\sqrt{D}$ and $y = y_1 + y_2\sqrt{D}$ gives a system of 2 equations, in the 4 variables x_1, x_2, y_1, y_2, for the variety $\mathrm{Res}_k^L(E)$:

$$y_1^2 + Dy_2^2 = x_1^3 + 3Dx_1x_2^2 + ax_1 + b,$$
$$2y_1y_2 = 3x_1^2x_2 + Dx_2^3 + ax_2.$$

Definitions and Properties of V_F

As before, L/k is a finite abelian extension and V is a commutative algebraic group over k. For every intermediate field F (i.e., $k \subseteq F \subseteq L$) such that F/k is cyclic, in Definition 2.4 below we will define a commutative algebraic group V_F over k such that, with $d := [F:k]$,

$$\operatorname{Res}_k^L(V) \text{ is isogenous over } k \text{ to } \bigoplus_{\substack{k \subseteq F \subseteq L \\ F/k \text{ cyclic}}} V_F \qquad (2.5)$$

via isogenies whose kernels are killed by [L:k],

$$V_F \text{ is isomorphic over } F \text{ to } V^{\varphi(d)} \qquad (2.6)$$

(so V_F is a twist of $V^{\varphi(d)}$ and

$$V_F(k) \cong \{\alpha \in V(F) : N_{F/M}(\alpha) = 1 \text{ for all } k \subseteq M \subsetneq F\}, \qquad (2.7)$$

where

$$N_{F/M}(\alpha) = \prod_{\sigma \in \operatorname{Gal}(F/M)} \sigma(\alpha) \in V(M)$$

and 1 is the identity element of V(k).

Taking k-points, it follows from (2.5) and (2.2) that there are homomorphisms (whose kernel and cokernel are "well-controlled" since those of the isogeny in (2.5) are) between the group V(L) and the direct sum $\oplus_F V_F(k)$. It follows from (2.6) that $\dim(V_F) = \varphi(d)\dim(V)$.

We next introduce notation needed for Definition 2.4. Let

$$G := \operatorname{Gal}(L/k).$$

If $g \in G$, then

$$\eta_{L/k}^g \in \mathrm{Hom}_L(\mathrm{Res}_k^L(V), V),$$

and by (2.1) there is a unique $gL/k, V \in \mathrm{End}_k(\mathrm{Res}_k^L(V))$ such that $\eta L/k \circ gL/k, V = \eta_{L/k}^g$. Extend $g \mapsto gL/k, V$ linearly to a ring homomorphism

$$\mathbb{Z}[G] \to \mathrm{End}_k(\mathrm{Res}_k^L(V)). \tag{2.8}$$

For $\alpha \in \mathbb{Z}[G]$, denote its image by

$$\alpha_{L/k,V} \in \mathrm{End}_k(\mathrm{Res}_k^L(V)).$$

On the level of k-points, if $\alpha = \sum_{g \in G} a_g\, g \in \mathbb{Z}[G]$, $v \in (\mathrm{Res}_k^L(V))(k)$, and $(\mathrm{Res}_k^L(V))(k)$ is identified with $V(L)$ as in (2.2), then

$$\alpha_{L/k,V}(v) = \prod_{g \in G} g(v)^{a_g}.$$

The map (2.8) is injective if the natural map $\mathbb{Z} \to \mathrm{End}_k(V)$ is injective. For example, the map (2.8) is injective when $V = \mathbb{G}_m$ or an abelian variety A, but is not when V is μ_n (the kernel of raising to the nth power on $\mathbb{G}_m$) or A[n] (the kernel of multiplication by n on A).

If $k \subseteq M \subseteq F$, let

$$N_{F/M} := \sum_{h \in \mathrm{Gal}(F/M)} h \in \mathbb{Z}[\mathrm{Gal}(F/M)] \subseteq \mathbb{Z}[\mathrm{Gal}(F/k)].$$

Summing the Gal(F/M)-conjugates of $\eta F/M$ gives a homomorphism

$$\sum_{h\in\text{Gal}(F/M)} \eta^h_{F/M} : \text{Res}^F_M(V) \to V$$

defined over M. Taking Res^M_k and using (2.3) gives a homomorphism

$$R_{F/M/k,V} : \text{Res}^F_k(V) = \text{Res}^M_k(\text{Res}^F_M(V)) \to \text{Res}^M_k(V) \qquad (2.9)$$

defined over k. on k-points, $R_{F/M/k, V}$ is the norm map from V(F) to V(M), which sends v to

$$\prod_{h\in\text{Gal}(F/M)} h(v).$$

There is a natural inclusion

$$\iota_{L/F/k,V} : \text{Res}^F_k(V) \hookrightarrow \text{Res}^L_k(V)$$

as follows. By (2.1) there is a homomorphism $\iota : V \to \text{Res}^L_F(V)$ de-fined over F such that $\eta L /F \circ \iota = \text{id}_V$ (on F-points, ι is the inclusion

$V(F) \subseteq V(L) \cong (\text{Res}^L_F(V))(F))$. The equation $\eta L /F \circ \iota = \text{id}_V$ shows that ι is injective. Applying Res^F_k and (2.3) gives the desired inclusion $\iota_{L /F/k, V}$. We will identify $\text{Res}^F_k(V)$ with its image in $\text{Res}^L_k(V)$. Note that

$$(N_{F/M})_{F/k,V} \in \text{End}_k(\text{Res}^F_k(V)), \qquad R_{F/M,k,V} \in \text{Hom}_k(\text{Res}^F_k(V), \text{Res}^M_k(V))$$
$$(N_{F/M})_{F/k,V} = \iota_{F/M/k,V} \circ R_{F/M/k,V}.$$

On k-points, $(N_{F/M})_{F/k,V}$ is the map from V (F) to V (F) that sends v to

$$\prod_{h\in\text{Gal}(F/M)} h(v).$$

Let

$$\Omega_{F/k} := \{\text{fields } M : k \subseteq M \subsetneq F\},$$
$$\Omega'_{F/k} := \{M \in \Omega_{F/k} : [F : M] \text{ is prime}\}.$$

Then every element of $\Omega_{F/k}$ is a subfield of some element of $\Omega'_{F/k}$. If d is a positive divisor of a positive integer n, let

$$\Psi_{n,d}(x) := \frac{x^n - 1}{\Phi_d(x)} \in \mathbb{Z}[x]$$

(Recall that $\Phi_d(x)$ is the dth cyclotomic polynomial) and let

$$\Psi_d(x) := \Psi_{d,d}(x) = \frac{x^d - 1}{\Phi_d(x)} \in \mathbb{Z}[x].$$

We next give several equivalent definitions for the variety

$$V_F \subseteq \mathrm{Res}_k^F(V) \subseteq \mathrm{Res}_k^L(V).$$

Definition 2.4

Suppose V is a commutative algebraic group over k, and F/k is cyclic of degree d. Fix a generator τ of $\mathrm{Gal}(F/k)$, and view $\Phi_d(\tau), \Psi_d(\tau) \in \mathbb{Z}[\mathrm{Gal}(F/k)]$ and $\Phi_d(\tau)_{F/k,V}, \Psi_d(\tau)_{F/k,V} \in \mathrm{End}_k(\mathrm{Res}_k^F(V))$. Then:

i. $V_F = \ker(\Phi_d(\tau)_{F/k,V}) \subseteq \mathrm{Res}_k^F(V),$

ii. $V_F = \bigcap_{M \in \Omega_{F/k}} \ker((N_{F/M})_{F/k,V}) = \bigcap_{M \in \Omega'_{F/k}} \ker((N_{F/M})_{F/k,V}) \subseteq \mathrm{Res}_k^F(V),$

iii. $V_F = \bigcap_{M \in \Omega_{F/k}} \ker(R_{F/M/k,V}) = \bigcap_{M \in \Omega'_{F/k}} \ker(R_{F/M/k,V}) \subseteq \mathrm{Res}_k^F(V),$

iv. if $F \subseteq L$, L/k is abelian of degree n, and $\sigma \in G := \mathrm{Gal}(L/k)$ is any element such that $\sigma|F$ generates $\mathrm{Gal}(F/k)$, then:

a. $V_F = (N_{L/F} \cdot \Psi_d(\sigma))_{L/k,v}(\text{Res}_k^L(V)) \subseteq \text{Res}_k^L(V),$

b. $V_F = \mathbb{Z}[G]_F \otimes_{\mathbb{Z}} V$ as defined in Sections 3 and 4.

(In (ii) and (iii) we adopt the convention that $V_k = V$.)

See [23] for the equivalence of the definitions; see also Proposition 5.1 below. Note that V_F is independent of the choice of L in (iv). Taking (iv) with L=F gives

$$V_F = (\Psi_d(\sigma))_{F/k,v}(\text{Res}_k^F(V)) \subseteq \text{Res}_k^F(V).$$

In [27], [28] and [29], the primitive subgroup V_F was defined as in (the first part of) Definition 2.4(iii) above. Taking k-points in (ii) or (iii) gives (2.7). When

$k = \mathbb{F}_q$, let $V_d := V_{\mathbb{F}_{q^d}}$.

Next, we continue the examples in Section 2.1.1 and Section 2.1.2. With notation D, L, G, σ, E as in Sections 2.1.1 and 2.1.2, let F=L. Then d=2, $N_{L/F} = 1, \Phi_d(\sigma) = \sigma + 1 = N_{L/k,}$, and $\Psi_d(\sigma) = \sigma - 1$.

$(\mathbb{G}_m)_L$ with Quadratic L

Let $V = \mathbb{G}_m$. We continue the example in Section 2.1.1. Recall that $\text{Res}_k^L(\mathbb{G}_m)$ is the variety R in Section 2.1.1. The image of σ under (2.8) is $\sigma_{L/k}, \mathbb{G}_m \in \text{End}_k(R)$ defined by $\sigma_{L/k'} \mathbb{G}_m (x_1, x_2, y) = (x_1, -x_2, y)$. Further,

$$\Phi_2(\sigma) = (N_{L/k})_{L/k,\mathbb{G}_m} = (\sigma + 1)_{L/k,\mathbb{G}_m} : R \to R$$

is given by

$$\left(x_1, x_2, \frac{1}{x_1^2 - Dx_2^2} \right) \mapsto \left(x_1^2 - Dx_2^2, 0, \frac{1}{(x_1^2 - Dx_2^2)^2} \right),$$

$R_{L/k/k', \mathbb{G}_m} : R \to \mathbb{G}_m$ is given by

$$\left(x_1, x_2, \frac{1}{x_1^2 - Dx_2^2} \right) \mapsto \left(x_1^2 - Dx_2^2, \frac{1}{x_1^2 - Dx_2^2} \right),$$

And

$$(N_{L/k} \cdot \Psi_2(\sigma))_{L/k, \mathbb{G}_m} = (\sigma - 1)_{L/k, \mathbb{G}_m} : R \to R$$

is given by

$$\left(x_1, x_2, \frac{1}{x_1^2 - Dx_2^2} \right) \mapsto \left(\frac{x_1^2 + Dx_2^2}{x_1^2 - Dx_2^2}, \frac{-2x_1 x_2}{x_1^2 - Dx_2^2}, 1 \right).$$

Then $(\mathbb{G}_m) L = (\sigma-1)_{L/k', \mathbb{G}_m} (R) = \ker((\sigma+1)_{L/k', \mathbb{G}_m})$, and $(\mathbb{G}_m)L$ is the subvariety of $R \subset \mathbb{A}^3$ defined by $x_1^2 - Dx_2^2 = 1 = y$. In particular, its k-points are the norm one elements of L. In [30], $(\mathbb{G}_m) L$ is called $\mathbb{T}_{L/k}$, and is called $\mathbb{T}_2$ in [29] and [30] when the fields are finite.

E_L with Elliptic Curve E and Quadratic L

We continue the example in Section 2.1.2. We saw that $\mathrm{Res}_k^L(E) = (E \times E^{(D)}) / T$. The image of σ under (2.8) is

$$\sigma_{L/k, E} \in \mathrm{End}((E \times E^{(D)})/T), \qquad \sigma_{L/k, E}(P, Q) = (P, -Q).$$

The natural inclusions of E and $E^{(D)}$ in $E \times E^{(D)}$ induce injective maps from E and E(D) into $(E \times E^{(D)})/T$. It is now easy to check that the image of $E^{(D)}$ in $(E \times E^{(D)})/T$ is both $\ker((\sigma+1)_{L/k, E})$ and $(\sigma-1)L/k, E((E \times E^{(D)})/T)$. Thus $EL = E^{(D)}$, by Definition 2.4(i), (ii) and (iii), or (iv)(a).

GENERAL CONSTRUCTIONS OF $\mathcal{I} \otimes_{\mathcal{O}} V$

In this section we state two definitions of a tensor product $\mathcal{I} \otimes_{\mathcal{O}} V$ that were given in [23], and give some examples. The varieties V_F in Definition 2.4 are special cases of $\mathcal{I} \otimes_{\mathcal{O}} V$, as we will discuss in Section 5. Theorem 3.2 states some useful properties of these tensor product varieties; in particular, Theorem 3.2(ii) motivates the notation $\mathcal{I} \otimes_{\mathcal{O}} V$.

Let k_s denote a separable closure of the field k, and let $G_k = \mathrm{Gal}(k_s / k)$.

From now on, suppose that V is commutative algebraic group over k, $\mathcal{O}$ is a commutative ring, $\mathcal{I}$ is a free $\mathcal{O}$-module of finite rank with a continuous right action of $G_{k'}$, and there is a ring homomorphism $\mathcal{O}$ $\rightarrow \mathrm{End}_k(V)$. We view $\mathcal{O}$ as a free rank one $\mathcal{O}$-module with trivial G_k-action. The reader can choose to restrict to the case $\mathcal{O} = \mathbb{Z}$ for simplicity; an example with $\mathcal{O} \neq \mathbb{Z}$ will appear only in Section 3.1.3.

Definition and Examples of $\mathcal{I} \otimes_{\mathcal{O}} V$

Definition 3.1: Let r be the rank of $\mathcal{I}$ as an $\mathcal{O}$-module, and fix an $\mathcal{O}$-module isomorphism $j : \mathcal{O}^r \xrightarrow{\;\sim\;} \mathcal{I}$. Let

$$c_{\mathcal{I}} \in H^1(k, \mathrm{Aut}_{k_s}(V^r))$$

be the image of the homomorphism $(\gamma \mapsto j^{-1} \circ j^{\gamma})$ under the composition

$$H^1(k, \mathrm{GL}_r(\mathcal{O})) \longrightarrow H^1(k, \mathrm{Aut}_k(V^r)) \longrightarrow H^1(k, \mathrm{Aut}_{k_s}(V^r))$$
$$\| \qquad\qquad\qquad\qquad \|$$
$$\mathrm{Hom}(G_k, \mathrm{GL}_r(\mathcal{O})) \qquad \mathrm{Hom}(G_k, \mathrm{Aut}_k(V^r))$$

induced by the homomorphism $\mathcal{O} \rightarrow \mathrm{End}_k(V)$. Let $\mathcal{I} \otimes_{\mathcal{O}} V$ be the twist of V^r by the cocycle $c_{\mathcal{I}}$, i.e., $\mathcal{I} \otimes_{\mathcal{O}} V$ is the unique commutative algebraic group over k with an isomorphism $\phi: V^r \xrightarrow{\sim} \mathcal{I} \otimes_O V$ defined over k_s such that for every $\gamma \in G_{k'}$

$$c_{\mathcal{I}}(\gamma) = \phi^{-1} \circ \phi^{\gamma}. \tag{3.1}$$

(See Corollaire to Proposition 5 on p. 131 in Section III-1.3 of [33], or Section 3.1 of [39], for twists of algebraic varieties.)

Note that the twists considered here and in [23] do not include all twists of V in the usual sense; that would require taking elements of $H^1(k_s, \mathrm{Aut}_{k_s}(V^r))$ rather than $H^1(k, \mathrm{Aut}_k(V^r))$.

Powers

Powers of V are a special case of $\mathcal{I} \otimes_O V$, namely, take $\mathcal{I} = \mathbb{Z}^r$ (with trivial Galois action), and let j be the identity map on $\mathbb{Z}^r$. Then the co-cycle $c_{\mathcal{I}}$ is trivial, and we can take ϕ to be the identity map on V^r, so $\mathbb{Z}^r \otimes_{\mathbb{Z}} V = V^r$. In particular, $V = \mathbb{Z} \otimes_{\mathbb{Z}} V$.

Restriction of Scalars

If L/k is a finite Galois extension with $G = \mathrm{Gal}(L/k)$, then (see Proposition 4.1 of [23])

$$\mathbb{Z}[G] \otimes_{\mathbb{Z}} V = \mathrm{Res}_k^L(V).$$

To see this, define $j: \mathbb{Z}^G \xrightarrow{\sim} \mathbb{Z}[G]$ by $(a_g)_{g \in G} \mapsto \sum_{g \in G} a_g g^{-1}$. By Definition 3.1, j induces an L-isomorphism $\phi: V^G \xrightarrow{\sim} \mathbb{Z}[G] \otimes_{\mathbb{Z}} V$. Composing ϕ^{-1} with the projection $V^G \to V$ onto the component corresponding to the identity element of G gives a homomorphism $\eta_{L/k}: \mathbb{Z}[G] \otimes_{\mathbb{Z}} V \to V$ that satisfies the universal property for $\mathrm{Res}_k^l(V)$.

Twists of Abelian Varieties

Let μ_n denote the group of nth roots of unity in $\overline{\mathbb{Q}}$, and let ζ_n denote a generator of μ_n.

Suppose A is an abelian variety over a field k and $\mu_n \to \text{End}(A)$. Suppose $\chi : G_k \to \mu_n$ is a homomorphism. Let $\mathcal{O} = \mathbb{Z}[\zeta_n] \to \text{End}(A)$, and let $\mathcal{I}$ be a free rank one $\mathbb{Z}[\zeta_n]$-module with G_k-action defined by $\alpha^\gamma = \chi(\gamma) \cdot \alpha$ for $\gamma \in G_k$ and $\alpha \in \mathbb{Z}[\zeta_n]$. Then the cocycle $c_{\mathcal{I}}$ of Definition 3.1 is χ^{-1}, and $\mathcal{I} \otimes_{\mathcal{O}} A$ is the twist of A by the character χ^{-1}. In Section 3.1.4, we give details in the case of quadratic twists of elliptic curves (see for example Section X.2 of [35]).

In Sections 3.1.4 and 3.1.5, use the notation in Sections 2.2.1, 2.2.2, 2.1.1 and 2.1.2. Then L/k is quadratic. Let

$$\chi_L : G_k \twoheadrightarrow \{\pm 1\}$$

be the quadratic character that factors through $G = \text{Gal}(L/k)$ and let $\mathcal{I}$ be a free rank one $\mathbb{Z}$-module with G_k-action defined by $a^\gamma = \chi_L(\gamma) \cdot a$ for $a \in \mathcal{I}$ and $\gamma \in G_k$. Fix a generator α of $\mathcal{I}$, and define $j : \mathbb{Z} \xrightarrow{\sim} \mathcal{I}$ by $n \mapsto n\alpha$. For $\gamma \in G_k$ and $n \in \mathbb{Z}$ we have $j^{-1} \circ j^\gamma(n) = \chi_L(\gamma) \cdot n$.

Quadratic Twists of Elliptic Curves

Continuing with the example in Sections 2.2.2 and 2.1.2, and using the above notation, we use Definition 3.1 with V=E to compute $\mathcal{I} \otimes_{\mathbb{Z}} E$. We have $c_{\mathcal{I}}(\gamma) = \chi_L(\gamma) = \phi^{-1} \circ \phi^\gamma$ with the isomorphism $\phi : E \xrightarrow{\sim} E^{(D)}$ of (2.4). By Definition 3.1, $\mathcal{I} \otimes_{\mathbb{Z}} E$ is $E^{(D)}$, the twist of E by the quadratic character χ_L.

Quadratic Twists of $\mathbb{G}_m$

Continuing with the example and notation in Sections 2.2.1 and 2.1.1, and using the above notation, we use Definition 3.1 with V=$\mathbb{G}_m$ to compute $\mathcal{I} \otimes_{\mathbb{Z}} \mathbb{G}_m$. Using the variety $(\mathbb{G}_m)_L$ of Section 2.2.1 and the isomorphism $\phi : \mathbb{G}_m \xrightarrow{\sim} (\mathbb{G}_m)_L$ defined by

$$(x, y) \mapsto \left(\frac{x+y}{2}, \frac{x-y}{2\sqrt{D}}, 1 \right)$$

(Whose inverse is the map $(a,b,1) \mapsto a+b\sqrt{D})$, we have $c_{\mathcal{I}}(\gamma) = \chi_L(\gamma) = \phi^{-1} \circ \phi^\gamma$. Thus, $\mathcal{I} \otimes_{\mathbb{Z}} \mathbb{G}_m$ is $(\mathbb{G}_m)_L$.

Properties of $\mathcal{I} \otimes_\mathcal{O} V$

The next result gathers together a number of results from [23].

Theorem 3.2

The variety $\mathcal{I} \otimes_\mathcal{O} V$ is a commutative algebraic group over k such that:

i. $\mathcal{I} \otimes_\mathcal{O} V$ is functorial in both V and $\mathcal{I}$.

ii. For all commutative k-algebras A and all Galois extensions F of k for which G_F acts trivially on $\mathcal{I}$,

$$(\mathcal{I} \otimes_\mathcal{O} V)(F \otimes_k A) \cong \mathcal{I} \otimes_\mathcal{O} (V(F \otimes_k A))$$

and

$$(\mathcal{I} \otimes_\mathcal{O} V)(A) \cong (\mathcal{I} \otimes_\mathcal{O} (V(F \otimes_k A)))^{\mathrm{Gal}(F/k)},$$

where the right-hand sides are the usual tensor products of $\mathcal{O}$-modules.

iii. If W is a commutative algebraic group over k and $\mathcal{J}$ is a free $\mathcal{O}$-module of finite rank with a continuous right action of $G_{k'}$ then there is a natural G_k-equivariant $\mathcal{O}$-module isomorphism

$$\text{Hom}_{\mathcal{O}}(\mathcal{I}, \mathcal{J}) \otimes_{\mathcal{O}} \text{Hom}_{k_s}(V, W) \xrightarrow{\sim} \text{Hom}_{k_s}(\mathcal{I} \otimes_{\mathcal{O}} V, \mathcal{J} \otimes_{\mathcal{O}} W) \quad (3.2)$$

that restricts to a homomorphism of $\mathcal{O}$-modules

$$\text{Hom}_{\mathcal{O}[G_k]}(\mathcal{I}, \mathcal{J}) \otimes_{\mathcal{O}} \text{Hom}_k(V, W) \hookrightarrow \text{Hom}_k(\mathcal{I} \otimes_{\mathcal{O}} V, \mathcal{J} \otimes_{\mathcal{O}} W).$$

iv. If F/k is separable, $\mathcal{J}$ is a free $\mathcal{O}$-module of finite rank with a continuous right action of G_k, and $\mathcal{I}$ and $\mathcal{J}$ are isomorphic as $\mathcal{O}[G_F]$-modules, then the commutative algebraic groups $\mathcal{I} \otimes_o V$ and $\mathcal{J} \otimes_o V$ are isomorphic over F.

v. If F/k is separable and G_F acts trivially on $\mathcal{I}$, then $\mathcal{I} \otimes_{\mathcal{O}} V$ is isomorphic over F to $V^{\text{rank}_{\mathcal{O}}(\mathcal{I})}$.

vi. If $0 \to \mathcal{I} \to \mathcal{J} \to \mathcal{K} \to 0$ is an exact sequence of free $\mathcal{O}$-modules of finite rank with a continuous right action of G_k, then the induced sequence $0 \to \mathcal{I} \otimes_o V \to \mathcal{J} \otimes_o V \to \mathcal{K} \otimes_o V \to 0$ is an exact sequence of commutative algebraic groups over k.

vii. If $\mathcal{I}, \mathcal{J}_1, \ldots, \mathcal{J}_t$ are free $\mathcal{O}$-modules of finite rank with a continuous right action of G_k, and $\mathcal{I} \otimes_{\mathbb{Z}} \mathbb{Q} \cong \otimes_{i=1}^{t}(\mathcal{J}_i \otimes_{\mathbb{Z}} \mathbb{Q})$ as $\mathcal{O}[G_k]$-modules, then

$$\mathcal{I} \otimes_{\mathcal{O}} V \text{ is } k\text{-isogenous to } \oplus_{i=1}^{t}(\mathcal{J}_i \otimes_{\mathcal{O}} V).$$

Proof:

See Lemma 1.3, Theorem 1.4, Corollary 1.7, Corollary 1.9, Theorem 2.1, Lemma 2.3, and Corollary 2.5 of [23]. Note that (vi) and (v) follow from (iv), which follows from (iii), which essentially follows from (i).

Theorem 3.2 can be used to show (2.5), (2.6) and (2.7) (see [23]). We show how this is done in Section 5, after defining $\mathbb{Z}[G]_F$. If $f \in \mathrm{Hom}_{\mathcal{O}}(\mathcal{I}, \mathcal{J})$ and $V = W$, let

$$f_V \in \mathrm{Hom}(\mathcal{I} \otimes_{\mathcal{O}} V, \mathcal{J} \otimes_{\mathcal{O}} V) \tag{3.3}$$

denote the image of $f \otimes \mathrm{id}_V$ under (3.2). When $\mathcal{O} = \mathbb{Z}$, $\mathcal{I} = \mathcal{J} = \mathbb{Z}[G]$, $V = W$, and $f \in \mathrm{End}_{\mathbb{Z}[G_k]}(\mathbb{Z}[G]) = \mathbb{Z}[G]$, then the map

$$\mathbb{Z}[G] \to \mathrm{End}_k(\mathrm{Res}_k^L(V)), \quad f \mapsto f_V \tag{3.4}$$

is the map (2.8).

An Alternate Definition of $\mathcal{I} \otimes_{\mathcal{O}} V$

As in the Appendix to [23], if $\mathcal{O}$ is a commutative noetherian ring, then even if the G_k-module and finitely generated $\mathcal{O}$-module $\mathcal{I}$ is not a free $\mathcal{O}$-module, one can define a tensor product $\mathcal{I} \otimes_{\mathcal{O}} V$ that coincides with the above definition where both make sense, as follows.

Definition 3.3

Take an $\mathcal{O}[G]$-presentation of $\mathcal{I}$, i.e., an exact sequence

$$\mathcal{O}[G]^a \xrightarrow{\psi} \mathcal{O}[G]^b \to \mathcal{I} \to 0 \tag{3.5}$$

of $\mathcal{O}[G]$-modules. Then

$$\mathcal{I} \otimes_{\mathcal{O}} V = \mathrm{coker}(\psi_V).$$

DECOMPOSITION OF GROUP RINGS

The decomposition of the restriction of scalars $\operatorname{Res}_k^L(V)$ into primitive subgroups arises from a decomposition of the group ring $\mathbb{Q}[\operatorname{Gal}(L/k)]$ into a direct sum of irreducible rational representations. See [32], especially exercise 13.1.

Suppose G is a finite abelian group. We will consider the group rings $\mathbb{Z}[G]$, $\mathbb{Q}[G]$, and $\mathbb{C}[G]$. Lemma 4.2 gives the decomposition of $\mathbb{Q}$ [G] and some properties of its constituent pieces $\mathbb{Z}[G]_H \otimes_{\mathbb{Z}} \mathbb{Q}$. In Section 5 we will see how to use Lemma 4.2 to obtain the properties of $\operatorname{Res}_k^L(V)$ and its constituent pieces V_F that were stated in Section 2.2.

We begin by decomposing $\mathbb{C}$ [G]. Let $\hat{G}$ be the character group of G, i.e., the set of homomorphisms from G to $\mathbb{C}^\times$. For $\chi \in \hat{G}$, let

$$e_\chi = \frac{1}{|G|} \sum_{g \in G} \chi(g) g^{-1} = \frac{1}{|G|} \sum_{g \in G} \chi^{-1}(g) g \in \mathbb{C}[G].$$

Then:

- $e_\chi^2 = e_\chi$,

- $e_\chi e_\psi = 0$ if $\chi \neq \psi$,

- $\sum_{\chi \in \hat{G}} e_\chi = 1$ (The identity element of G),

- $e_\chi \mathbb{C}[G] = e_\chi \mathbb{C}$, a one-dimensional $\mathbb{C}$-vector space,

- $\mathbb{C}[G] = \sum_{\chi \in \hat{G}} (e_\chi \cdot \mathbb{C}[G]) = \bigoplus_{\chi \in \hat{G}} (e_\chi \cdot \mathbb{C}[G]) = \bigoplus_{\chi \in \hat{G}} e_\chi \mathbb{C}.$

A problem with decomposing $\mathbb{Q}[G]$ or $\mathbb{Z}[G]$ is that $\chi(g)$ is not necessarily in $\mathbb{Q}$, so the idempotents e_χ are not necessarily in $\mathbb{Q}[G]$. One therefore needs to consider a sum of e_χ's, corresponding to an irreducible rational representation of G.

Lemma 4.1

Let C_G be the set of subgroups H of G such that G/H is cyclic, let R_G be the set of irreducible rational representations of G, and let X_G be the set of $G_\mathbb{Q}$-orbits of $\hat{G}$, where $G_\mathbb{Q} := \mathrm{Gal}(\bar{\mathbb{Q}}/\mathbb{Q})$. Then C_G, R_G, and X_G are in natural one-to-one correspondence.

Proof:

If $H \in C_G$, let $Y_H : \{\chi \in \hat{G} : \ker(\chi) = H\} \in X_G$. Conversely, if $Y \in X_G$, let $H_Y := \ker(\chi)$ for any $\chi \in Y$; then $G/H_Y \cong \chi(G)$, a finite, and therefore cyclic, subgroup of $\mathbb{C}^\times$, so $H_Y \in C_G$. If $Y \in X_G$, then $\sum_{\chi \in Y} e_\chi \in \mathbb{Q}[G]$, and the action of G on $\sum_{\chi \in Y}(e_\chi)\mathbb{Q}[G]$ is an irreducible rational representation ρY of G, so $\rho Y \in R_G$. Conversely, if $\rho \in R_G$, decompose ρ over $\mathbb{C}$ into a direct sum of characters of G. Since ρ is rational and irreducible, this gives a single $G_\mathbb{Q}$-orbit Y_ρ of $\hat{G}$. If $H \in C_G$, let

$$e_H = \sum_{\chi \in Y_H} e_\chi \in \mathbb{Q}[G].$$

Then:

- $e_H^2 = e_H,$

- $e_{H_1} e_{H_2} = 0$ if $H_1 \neq H_2$, and

• $\sum_{H \in C_G} e_H = 1.$

Let $\mathbb{Q}[G]_H = e_H \cdot \mathbb{Q}[G]$, a simple $\mathbb{Q}[G]$-submodule of $\mathbb{Q}[G]$. Then $\mathbb{Q}[G]_H$ is the unique irreducible rational representation of G contained in $\mathbb{Q}[G]$ whose kernel is H, and

$$\mathbb{Q}[G] = \bigoplus_{H \in C_G} (e_H \cdot \mathbb{Q}[G]) = \bigoplus_{H \in C_G} \mathbb{Q}[G]_H. \tag{4.1}$$

Let

$$\mathbb{Z}[G]_H = \mathbb{Q}[G]_H \cap \mathbb{Z}[G]$$

and let

$$N_H = \sum_{h \in H} h.$$

Since $\mathbb{Z}[G]_H$ is a submodule of $\mathbb{Z}[G]$, it is a free $\mathbb{Z}$-module.

Lemma 4.2

Suppose G is a finite abelian group, $H \in C_G$, $\sigma \in G$ is such that σH is a generator of G/H, and $d := |G/H|$. Then:

i. $\mathbb{Z}[G]_H = N_H \cdot \Psi_d(\sigma) \cdot \mathbb{Z}[G] \cong \mathbb{Z}[x]/(\Phi_d(x))$,

ii. $\mathbb{Z}[G]_H \otimes_{\mathbb{Z}} \mathbb{Q} = N_H \cdot \Psi_d(\sigma) \cdot \mathbb{Q}[G] = \mathbb{Q}[G]_H$,

iii. $\mathrm{rank}_{\mathbb{Z}}(\mathbb{Z}[G]_H) = \varphi(d)$,

iv. $\mathbb{Z}[G] \otimes_{\mathbb{Z}} \mathbb{Q} = \mathbb{Q}[G] = \bigoplus_{H \in C_G} (\mathbb{Z}[G]_H \otimes_{\mathbb{Z}} \mathbb{Q})$,

v. $\mathbb{Z}[G]/\bigoplus_{H\in C_G}\mathbb{Z}[G]_H$ is killed by $|G|$.

vi. If further G is cyclic of order n and σ generates G, then viewing $\Phi_d(\sigma) \in \mathrm{End}(\mathbb{Z}[G])$ we have

$$\mathbb{Z}[G]_H = \Psi_{n,d}(\sigma) \cdot \mathbb{Z}[G] = \ker(\Phi_d(\sigma)).$$

Proof:

Let $\beta = N_H \cdot \Psi_d|(\sigma)$. For $\chi \in \hat{G}$, if $\ker(\chi)=H$ then $e_\chi \cdot \beta \cdot \mathbb{C}[G]=e_\chi \cdot \mathbb{C}=e_\chi \cdot e_H \cdot \mathbb{C}[G]$, while if $\ker(\chi)\neq H$ then $e_\chi \cdot \beta \cdot \mathbb{C}[G]=0=e_\chi \cdot e_H \cdot \mathbb{C}[G]$. It follows that $\beta \cdot \mathbb{C}[G]=e_H \cdot \mathbb{C}[G]$. By linear algebra,

$$\beta \cdot \mathbb{Q}[G] = e_H \cdot \mathbb{Q}[G] = \mathbb{Q}[G]_H. \tag{4.2}$$

By inspection, $N_H\mathbb{Q}[G]\cap\mathbb{Z}[G] = N_H\mathbb{Z}[G]$, and it follows that $\mathbb{Z}[G]/N_H\mathbb{Z}[G]$ is a torsion-free $\mathbb{Z}$-module. The map

$$\pi_H : N_H\mathbb{Z}[G] \xrightarrow{\sim} \mathbb{Z}[G/H], \qquad N_H\sum_{g\in G}a_g g \mapsto \sum_{g\in G}a_g(gH)$$

is an isomorphism of $\mathbb{Z}[G]$-modules that induces an isomorphism

$$N_H\mathbb{Z}[G]/\beta\mathbb{Z}[G] \xrightarrow{\sim} \mathbb{Z}[G/H]/\Psi_d(\sigma H)\mathbb{Z}[G/H] \cong \mathbb{Z}[x]/(\Psi_d(x)),$$

and the latter is a torsion-free $\mathbb{Z}$-module since $\Psi_d(x)$ is monic. From the exact sequence

$$0 \longrightarrow N_H\mathbb{Z}[G]/\beta\mathbb{Z}[G] \longrightarrow \mathbb{Z}[G]/\beta\mathbb{Z}[G] \longrightarrow \mathbb{Z}[G]/N_H\mathbb{Z}[G] \longrightarrow 0$$

it follows that $\mathbb{Z}[G]/\beta\mathbb{Z}[G]$ is a torsion-free $\mathbb{Z}$-module, and now (4.2) gives that $\beta \cdot \mathbb{Z}[G]=\mathbb{Z}[G]_H$. Further, via π_H,

$$\beta\mathbb{Z}[G] \xrightarrow{\sim} \Psi_d(\sigma H)\mathbb{Z}[G/H] \xrightarrow{\sim} \Psi_d(x)(\mathbb{Z}[x]/(x^d-1)) \cong \mathbb{Z}[x]/(\Phi_d(x)),$$

and we have (i) and (ii). Since $\mathrm{rank}_{\mathbb{Z}}(\mathbb{Z}[x]/(\Phi_d(x))) = \varphi(d)$, we have (iii). Now (iv) follows from (4.1) and (ii). If $\alpha \in \mathbb{Z}[G]$, then

$$|G| \cdot \alpha = \sum_{H \in C_G} e_H |G| \alpha \in \bigoplus_{H \in C_G} \mathbb{Z}[G]_H.$$

Thus

$$|G| \cdot \mathbb{Z}[G] \subseteq \bigoplus_{H \in C_G} \mathbb{Z}[G]_H \subseteq \mathbb{Z}[G] \tag{4.3}$$

and (v) follows.

Suppose G is cyclic of order n with generator σ, and view $\mathbb{Z}[x]/(x^n-1)$ as a G-module with σ acting as multiplication by x. Since

$$\Psi_{n,d}(x) = (1 + x^d + x^{2d} + \cdots + x^{n-d})\Psi_d(x)$$

we have $\Psi_{n,d}(\sigma) = N_H \Psi_d(\sigma)$. By (i), we have $\mathbb{Z}[G]_H = \Psi_{n,d}(\sigma)\mathbb{Z}[G]$. Since the latter is isomorphic to $\Psi_{n,d}(x)(\mathbb{Z}[x]/(x^n-1))$, which is the kernel of multiplication by $\Phi_d(x)$ in $\mathbb{Z}[x]/(x^n-1)$, it follows that $\Psi_{n,d}(\sigma)\mathbb{Z}[G]$ is the kernel of multiplication by $\Phi_d(\sigma)$ in $\mathbb{Z}[G]$.

PRIMITIVE SUBGROUPS, REVISITED

Suppose that L/k is a finite abelian extension, $G = \mathrm{Gal}(L/K), k \subseteq F \subseteq L, F/k, F/k$ is cyclic, $d=[F:k]$, and $H = \mathrm{Gal}(L/F)$. Let $\mathbb{Z}[G]_F := \mathbb{Z}[G]_H$ and $\mathbb{Q}[G]_F := \mathbb{Q}[G]_H$.

Suppose that V is a commutative algebraic group over k. Viewing $\mathbb{Z}[G]_F$ as a G_k-module, then $\mathbb{Z}[G]_{F \otimes_{\mathbb{Z}} k} V$ (as defined using Definition 3.1) is a commutative algebraic group over k.

Suppose $k \subseteq M \subseteq F$. Letting

$$R_{F/M/k} : \mathbb{Z}[\mathrm{Gal}(F/k)] \to \mathbb{Z}[\mathrm{Gal}(M/k)]$$

be the natural projection map, then the map

$$R_{F/M/k,V} \in \mathrm{Hom}_k(\mathrm{Res}_k^F(V), \mathrm{Res}_k^M(V))$$

defined in (2.9) is the same as the map $(R_{F/M/k})_V$ obtained from $R_{F/M/k}$ via (3.3) (with $\mathcal{O} = \mathbb{Z}, \mathcal{I} = \mathbb{Z}[\mathrm{Gal}(F/k)]$, and $J = \mathbb{Z}[\mathrm{Gal}(M/K)]$).

Proposition 5.1

With notation as in Definition 2.4, parts (a) and (b) of Definition 2.4 (iv) are equivalent, i.e.,

$$\mathbb{Z}[G]_F \otimes_{\mathbb{Z}} V = \beta_{L/k,V}(\mathrm{Res}_k^L(V)),$$

where $\beta_{L/k,V} = N_{L/F} \cdot \Psi_d(\sigma) \in \mathbb{Z}[G]$.

Proof:

Lemma 4.2 (i) gives a diagram

$$0 \longrightarrow \ker(\beta) \longrightarrow \mathbb{Z}[G] \overset{\beta}{\longrightarrow} \mathbb{Z}[G]_F \longrightarrow 0.$$

with a diagonal map β from $\mathbb{Z}[G]$ to $\mathbb{Z}[G]$ and a vertical map into $\mathbb{Z}[G]$.

Since $\ker(\beta) \subseteq \mathbb{Z}[G]$, $\ker(\beta)$ is torsion-free, and is thus a free $\mathbb{Z}$-module. By Theorem 3.2(vi), there is an induced diagram

$$0 \longrightarrow \ker(\beta) \otimes_{\mathbb{Z}} V \longrightarrow \mathrm{Res}_k^L(V) \xrightarrow{\beta_V} \mathbb{Z}[G]_F \otimes_{\mathbb{Z}} V \longrightarrow 0$$

$$\mathrm{Res}_k^L(V)$$

which shows that $\mathbb{Z}[G]_F \otimes_{\mathbb{Z}} V = \beta_V(\mathrm{Res}_k^L(V)) = \beta_{L/k,V}(\mathrm{Res}_k^L(V)).$

We can now show (2.6) and (2.5).

Proposition 5.2

V_F is isomorphic over F to $V^{\phi(d)}$.

Proof:

Since $\mathbb{Z}[G]_F$ is a free $\mathbb{Z}$-module of rank $\phi(d)$ and G_F acts trivially on $\mathbb{Q}[G]_F$, we have $\mathbb{Z}[G]_F \cong \mathbb{Z}^{\phi(d)}$ as $\mathbb{Z}[G_F]$-modules. The result now follows from Theorem 3.2(v).

Proposition 5.3

The algebraic varieties $\mathrm{Res}_k^L(V)$ and $\bigoplus_{\substack{k \subseteq F \subseteq L \\ F/k \, \mathrm{cyclic}}} V_F$ are k-isogenous, via isogenies whose kernels are killed by $|G|$.

Proof:

Apply Lemma 4.2(iv) and (v) and Theorem 3.2(vii) with $\mathcal{O} = \mathbb{Z}, \mathcal{I} = \mathbb{Z}[G]$, and $\{\mathcal{J}_i\} = \{\mathbb{Z}[G]_F\}$. The inclusions (4.3) induce a sequence of isogenies

$$\mathrm{Res}_k^L(V) \to \bigoplus_{\substack{k \subseteq F \subseteq L \\ F/k \, \mathrm{cyclic}}} V_F \to \mathrm{Res}_k^L(V)$$

whose composition is raising to the power $|G|$.

Suppose G is generated by $\sigma, g_2,\ldots,g_a$, where $g_2,\ldots,g_a \in H$. Then (3.5) with $\mathcal{O} = \mathbb{Z}$ and $\mathcal{I} = \mathbb{Z}[G]_F$ can be taken to be $\mathbb{Z}[G]^a \to \mathbb{Z}[G] \to \mathbb{Z}[G]_F \to 0$,

where the first map is defined by $(\alpha_1,\ldots,\alpha_a) \mapsto \alpha_1 \Phi_d(\sigma) + \sum_{i=2}^{a} \alpha_i(g_i - 1)$ and the second is multiplication by $N_{L/F}\Psi_d(\sigma)$.

Note that $\Psi_1(x) = 1$ and $\phi_1(x) = x - 1$, so $\mathbb{Z}[G]_k = N_{L/k} \cdot \mathbb{Z}$, a free rank one $\mathbb{Z}$-module with trivial Galois action, and

$$V_k = V = \mathbb{Z}[G]_k \otimes_{\mathbb{Z}} V = \ker((0)_{k/k,V}) = (N_{L/k})_V(\mathrm{Res}_k^L(V)).$$

In the following subsections we give some special cases of primitive subgroups.

Trace Zero Subgroups

Suppose $d = [F:k]$ is prime. Then (by Definition 2.4(ii) or (iii)) V_F is the norm one subgroup of $\mathrm{Res}_k^F(V)$ if the group law on V is viewed multiplicatively, and is the trace zero subgroup when the group law is viewed additively. Further, $\mathrm{Res}_k^F(V)$ is k-isogenous to $V \times V_F$.

Trace zero subgroups of the restriction of scalars for abelian varieties appear in [8] and [19].

Decomposition of $\mathrm{Res}_k^L(\mathbb{G}_m)$

Continuing the example in Section 2.1.1, where L/k is quadratic and $\mathrm{Res}_k^L(\mathbb{G}_m) = R \subset \mathbb{A}^3$, we can give the decomposition of $\mathrm{Res}_k^L(\mathbb{G}_m)$ into $\mathbb{G}_m \times (\mathbb{G}_m)_L$ (up to isogeny) explicitly. The homomorphism

$$\mathbb{G}_m \times (\mathbb{G}_m)_L \to \mathrm{Res}_k^L(\mathbb{G}_m), \qquad ((x,y),(a,b,1)) \mapsto (xa, xb, y^2)$$

has kernel $\{((1,1),(1,0,1)),((-1,-1),(-1,0,1))\}$ of order 2. Composing, in either order, with the homomorphism gives the squaring map.

$$\mathrm{Res}_k^L(\mathbb{G}_m) \to \mathbb{G}_m \times (\mathbb{G}_m)_L, \qquad (x_1, x_2, y) \mapsto ((y^{-1}, y), ((x_1^2 + Dx_2^2)y, 2x_1x_2y, 1))$$

Quadratic Twists of Elliptic Curves

Quadratic twists of elliptic curves are examples of both twists of abelian varieties and primitive (in fact, trace zero) subgroups. We continue with the example and notation of Section 3.1.4. Then

$$\mathbb{Z}[G]_k = N_{L/k}\,\Psi_1(\sigma)\mathbb{Z}[G] = (\sigma + 1)\mathbb{Z} = e_G\mathbb{Q}[G] \cap \mathbb{Z}[G],$$
$$\mathbb{Z}[G]_L = \Psi_2(\sigma)\mathbb{Z}[G] = (\sigma - 1)\mathbb{Z} = e_{\{1\}}\mathbb{Q}[G] \cap \mathbb{Z}[G],$$

free rank one $\mathbb{Z}$-modules, and G_k-modules with $\gamma \in G_k$ acting on $\mathbb{Z}[G]_L$ (resp., $\mathbb{Z}[G]_k$) as multiplication by $\chi_L(\gamma)\in\{\pm1\}$ (resp., trivially). (Note that $e_G = ex_0$ with χ_0 the trivial character, and $e\{1\} = ex_1$ with χ_1 $(\sigma) = -1$.). We saw in Section 3.1.4 that $\mathbb{Z}[G]_L \otimes \mathbb{Z} E = E^{(D)}$. Similarly, $\mathbb{Z}[G]_k \otimes \mathbb{Z} E = E.$.

Next we check that Definition 3.3 gives the same answer for $\mathbb{Z}[G]_L \otimes \mathbb{Z} E.$. Consider the presentation

$$\mathbb{Z}[G] \xrightarrow{\sigma+1} \mathbb{Z}[G] \xrightarrow{\sigma-1} \mathbb{Z}[G]_L \longrightarrow 0.$$

Since the sequence

$$(E \times E^{(D)})/T \xrightarrow{(\sigma+1)_E} (E \times E^{(D)})/T \xrightarrow{(\sigma-1)_E} E^{(D)} \longrightarrow 0$$

is exact, there is a natural identification of $E^{(D)}$ with $\mathrm{coker}^{((\sigma+1)_E)}$, as desired.

Similarly, considering

$$(E \times E^{(D)})/T \xrightarrow{(\sigma-1)_E} (E \times E^{(D)})/T \xrightarrow{(\sigma+1)_E} E \longrightarrow 0$$

identifies E with coker $^{((\sigma + 1)_E)}$, as desired. To summarize,

$$E_k = E = \ker((\sigma - 1)_E) = \operatorname{coker}((\sigma - 1)_E)$$
$$= (\sigma + 1)_E((E \times E^{(D)})/T) = \mathbb{Z}[G]_k \otimes_{\mathbb{Z}} E,$$
$$E_L = E^{(D)} = \ker((\sigma + 1)_E) = \operatorname{coker}((\sigma + 1)_E)$$
$$= (\sigma - 1)_E((E \times E^{(D)})/T) = \mathbb{Z}[G]_L \otimes_{\mathbb{Z}} E.$$

The maps $\mathbb{Z}[G]_k \times \mathbb{Z}[G]_L \hookrightarrow \mathbb{Z}[G] \to \mathbb{Z}[G]_k \times \mathbb{Z}[G]_L$, where the right-hand map is $\alpha \mapsto ((1 + \sigma)\alpha, (1 - \sigma)\alpha)$ so the composition is multiplication by 2, induce isogenies

$$E_k \times E_L = E \times E^{(D)} \twoheadrightarrow \operatorname{Res}_k^L(E) = (E \times E^{(D)})/T \twoheadrightarrow E \times E^{(D)},$$

where the left isogeny is the natural quotient map and the right isogeny and the composition are multiplication by 2.

Algebraic Tori

In this section, $V = \mathbb{G}_m$.

Proposition 5.4

Suppose $k = \mathbb{F}_q$, $L = \mathbb{F}_{q^n}$, d is a divisor of n, and $F = \mathbb{F}_{q^d}$. Then:

i. $(\mathbb{G}_m)_F(k) \subseteq F^{\times}$,

ii. the group $(\mathbb{G}m)_F(k)$ is isomorphic to the subgroup of $F^{\times}$ of order $\Phi_d(q)$,

iii. if $v \in (\mathbb{GF}_m)_F(k) \subseteq F^{\times}$ and v has prime order not dividing d, then for all proper intermediate fields M(i.e., $k \subseteq M \subsetneq F$), we have $v \notin M$.

Proof:

Part (i) follows from Definition 2.4. If $\sigma \in G = \text{Gal}(\mathbb{F}_{q^n} / \mathbb{F}_q)$ is the map $x \mapsto x^q$, then the map

$$\mathbb{Z}[G] \rightarrow \text{End}_k(\text{Res}_k^L(\mathbb{G}_m))$$

of (3.4) or (2.8) sends $\sum_{i=0}^{n-1} a_i \sigma^i$ to the map that takes v to $v^{\sum a_i q^i}$. Further,

$$(\mathbb{G}_m)_F(\mathbb{F}_q) = \ker(\Phi_d(\sigma)_{L/k,\mathbb{G}_m}) = \ker(v \mapsto v^{\Phi_d(q)}),$$

Which is the subgroup of $\mathbb{F}_{q^d}^\times$ (and of $\mathbb{F}_{q^n}^\times$) of order $\Phi_d(q)$. For (iii), see Lemma 1 of [1].

Note that if $\mathcal{I}$ is a free $\mathbb{Z}$-module of finite rank with a continuous right action of G_k, then $\mathcal{I} \otimes_{\mathbb{Z}} \mathbb{G}_m$ is the algebraic torus whose character module $\text{Hom}(\mathcal{I} \otimes_{\mathbb{Z}} \mathbb{G}_m, \mathbb{G}_m)$ is $\text{Hom}(\mathcal{I}, \mathbb{Z})$ (see Example 6 in Section 3.4 of [39] or Corollary 1.10 of [23]).

Algebraic Tori over Finite Fields

The primitive subgroup $(\mathbb{G}_m)_{\mathbb{F}_{q^d}}$ was denoted $\mathbb{T}_d$ in [27], [28], [29] and [30].

Write $(\mathbb{G}_m)_d := (\mathbb{G}_m)_F$ with $F = \mathbb{F}_{q^d}$. By (2.5) and (2.2), $\mathbb{F}_{q^n}^\times$ can be viewed as "almost isomorphic" to $\bigoplus_{d|n}(\mathbb{G}_m)_d(\mathbb{F}_q)$, and therefore cryptography in $\mathbb{F}_{q^n}^\times$ can be reduced to cryptography in primitive subgroups $(\mathbb{G}_m)_d(\mathbb{F}_q)$ for the divisors d of n. Proposition 5.4(i) implies that attacks (e.g., index calculus attacks) on the discrete log problem in $\mathbb{F}_{q^n}^\times$ give attacks on the discrete log problem in $(\mathbb{G}_m)_d(\mathbb{F}_q)$. Proposition 5.4(iii) tells us that to attack the primitive subgroup $(\mathbb{G}_m)_d(\mathbb{F}_q)$ via an attack that requires using the full multiplicative group of a finite field, no proper subfield of suffices. It is in this sense that the subgroup $(\mathbb{G}_m)_n(\mathbb{F}_q)$ of $\mathbb{F}_{q^n}^\times$ $(=\mathbb{G}_m(\mathbb{F}_{q^n}))$ of order $\Phi_n(q)$ is often thought of as the most cryptographically secure primitive subgroup. More generally, one can argue that when V is a commutative algebraic group over $\mathbb{F}_q$, then it makes sense to think of $V_n(\mathbb{F}_q)$ as the most cryptographically secure primitive subgroup of $V(\mathbb{F}_{q^n})$.

An Example with n=6

Suppose G is a cyclic group of order 6. For d=2, 3, let H_d denote the subgroup of G of index d (and order 6/d). Then

$$\mathbb{Q}[G] = \mathbb{Q}[G]_G \oplus \mathbb{Q}[G]_{H_2} \oplus \mathbb{Q}[G]_{H_3} \oplus \mathbb{Q}[G]_1,$$
$$\dim_\mathbb{Q}(\mathbb{Q}[G]_G) = \varphi(1) = 1 = \varphi(2) = \dim_\mathbb{Q}(\mathbb{Q}[G]_{H_2}),$$
$$\dim_\mathbb{Q}(\mathbb{Q}[G]_{H_3}) = \varphi(3) = 2 = \varphi(6) = \dim_\mathbb{Q}(\mathbb{Q}[G]_1).$$

Let $\zeta_6 = (1+\sqrt{-3})/2$ and $\zeta_3 = \zeta_6^2 = (-1+\sqrt{-3})/2$. Then

$$\mathbb{Q}[G]_G = e_G \cdot \mathbb{Q} = e_G \cdot \mathbb{Q}[G] = \Psi_{6,1}(\sigma)\mathbb{Q}[G],$$

where

$$e_G = \tfrac{1}{6}(1 + \sigma + \sigma^2 + \sigma^3 + \sigma^4 + \sigma^5) = \tfrac{1}{6}\Psi_{6,1}(\sigma);$$
$$\mathbb{Q}[G]_{H_2} = e_{H_2} \cdot \mathbb{Q} = e_{H_2} \cdot \mathbb{Q}[G] = \Psi_{6,2}(\sigma)\mathbb{Q}[G],$$

where

$$e_{H_2} = \tfrac{1}{6}(1 - \sigma + \sigma^2 - \sigma^3 + \sigma^4 - \sigma^5) = -\tfrac{1}{6}\Psi_{6,2}(\sigma);$$
$$\mathbb{Q}[G]_{H_3} = e_{H_3} \cdot \mathbb{Q}[G] = \Psi_{6,3}(\sigma)\mathbb{Q}[G],$$

where

$$e_{H_3} = e_{\chi_3} + e_{\chi_3^2} =$$

$$\frac{1}{6}[(1 + \zeta_3\sigma + \zeta_3^2\sigma^2 + \sigma^3 + \zeta_3\sigma^4 + \zeta_3^2\sigma^5) + (1 + \zeta_3^2\sigma + \zeta_3\sigma^2 + \sigma^3 + \zeta_3^2\sigma^4 + \zeta_3\sigma^5)]$$

$$= \frac{1}{6}(2 - \sigma - \sigma^2 + 2\sigma^3 - \sigma^4 - \sigma^5) = -\frac{1}{6}\Psi_{6,3}(\sigma)(\sigma + 2);$$

$$\mathbb{Q}[G]_1 = e_1 \cdot \mathbb{Q}[G] = \Psi_{6,6}(\sigma)\mathbb{Q}[G].$$

Where

$$e_1 = e_{\chi_6} + e_{\chi_6^{-1}} =$$

$$\frac{1}{6}[(1 + \zeta_6\sigma + \zeta_6^2\sigma^2 - \sigma^3 - \zeta_6\sigma^4 - \zeta_6^2\sigma^5) + (1 + \zeta_6^{-1}\sigma + \zeta_6^{-2}\sigma^2 - \sigma^3 - \zeta_6^{-1}\sigma^4 - \zeta_6^{-2}\sigma^5)]$$

$$= \frac{1}{6}(2 + \sigma - \sigma^2 - 2\sigma^3 - \sigma^4 + \sigma^5) = \frac{1}{6}\Psi_{6,6}(\sigma)(\sigma - 2).$$

Then

$$\mathbb{Z}[G]_G = \Psi_{6,1}(\sigma)\mathbb{Z}[G] = (1 + \sigma + \sigma^2 + \sigma^3 + \sigma^4 + \sigma^5)\mathbb{Z}[G] = \ker(\sigma - 1),$$
$$\mathbb{Z}[G]_{H_2} = \Psi_{6,2}(\sigma)\mathbb{Z}[G] = (-1 + \sigma - \sigma^2 + \sigma^3 - \sigma^4 + \sigma^5)\mathbb{Z}[G] = \ker(\sigma + 1),$$
$$\mathbb{Z}[G]_{H_3} = \Psi_{6,3}(\sigma)\mathbb{Z}[G] = (-1 + \sigma - \sigma^3 + \sigma^4)\mathbb{Z}[G] = \ker(\sigma^2 + \sigma + 1),$$
$$\mathbb{Z}[G]_{\{1\}} = \Psi_{6,6}(\sigma)\mathbb{Z}[G] = (-1 - \sigma + \sigma^3 + \sigma^4)\mathbb{Z}[G] = \ker(\sigma^2 - \sigma + 1).$$

(The fact that $\sigma \pm 2$ is invertible in $\mathbb{Q}[G]$ follows from the fact that, after extending χ to a ring homomorphism $\chi: \mathbb{C}[G] \to \mathbb{C}$, then $\alpha \in \mathbb{Q}[G]$ is a unit if and only if $\chi(\alpha) \neq 0$ for all $\hat{G}$.)

If V is a commutative algebraic group over a field k and L is a cyclic degree 6 extension of k, for d=2, 3 let F_d denote the degree d extension of k in L and let $G = \text{Gal}(L/k)$. Then

$$\mathbb{Z}[G]_{F_d} = \mathbb{Z}[G]_{H_d} = \ker(N_{F_d/k}),$$
$$\mathbb{Z}[G]_L = \mathbb{Z}[G]_{\{1\}} = \ker(N_{L/F_2}) \cap \ker(N_{L/F_3}),$$

$V_k = V$, and V_{F_2} is the quadratic twist of V with respect to F_2.

If $k=\mathbb{F}_q$, then $V_L=\ker(q^2-q+1)=\ker(\Phi_6(q))$, the subgroup of $\mathbb{F}_{q^6}^x$ of order $\Phi_6(q)$. The cryptosystem CEILIDH [27] is based on this variety, while XTR is based on a quotient of V_L by an action of the symmetric group S_3 (see [27], [28], [29] and [30]).

OPEN QUESTIONS AND FUTURE DIRECTIONS

To do efficient discrete log cryptography in $V(\mathbb{F}_{q^n})$, where V is a commutative algebraic group over $\mathbb{F}_q$, we saw above that one can reduce to considering the subgroups $V_d(\mathbb{F}_q)$, where d is a divisor of n and V_d is V_F with It makes sense to think of V_n as the most cryptographically secure of the subvarieties V_d, thereby reducing to the case of V_n. Since dim $(V_n)=\varphi(n)$ dim (V), in the case where dim $(V)=1$ (let us restrict to that case), to obtain the greatest efficiency one would like to represent the elements of $V_n(\mathbb{F}_q)$ using only $\phi(n)$ elements of $\mathbb{F}_q$. In other words, one would like a low degree compression map $V_n -- \to \mathbb{A}^{\varphi(n)}$ defined over $\mathbb{F}_q$ where $\mathbb{A}^r$ is affine space of dimension r, along with an efficiently computable decompression function. (We allow the compression and decompression maps to be rational maps, defined only on a Zariski open subset.)

This is done in [26] when V is an elliptic curve E and n=3 or 5, with morphisms $E_n-\{0\} \to \mathbb{A}^{\varphi(n)}$ of degree 8 and 54, respectively (for n=2 or 1, it is easy to do with a degree 2 map). For larger primes n, it is not known how to efficiently decompress elements of $\mathbb{F}_q^{\varphi(n)}$ (the method in [26] and [34] gives a compression function for which the degree has not been computed, and for which no efficient decompression algorithm is known).

When $V=\mathbb{G}_m$, a trace map is used to give a degree 2 morphism $(\mathbb{G}_m)_n \to \mathbb{A}^{\varphi(n)}$ when n=2 in [24], [36], [37], [41] and [42] and a degree 6 map when n=6 in [2] and [20] (for a degree 3 map using 2 symmetric functions when n=3, see [13]). When $V = \mathbb{G}_m$, a degree 1 map (birational isomorphism) $(\mathbb{G}_m)_n \dashrightarrow \mathbb{A}^{\varphi(n)}$ is given in [27] when n=2 or 6. Explicitly (see [27] and [29]), for n=2 we have

$$(\mathbb{G}_m)_2 \dashrightarrow \mathbb{A}^1, \qquad (a, b, 1) \mapsto \frac{1+a}{b},$$

with inverse map $\alpha \mapsto (\frac{\alpha^2 + D}{\alpha^2 - D}, \frac{2\alpha}{\alpha^2 - D}, 1)$. Via this map, the group law on $(\mathbb{G}_m)_2$ induces an operation $\alpha * \beta = (\alpha\beta + D)/(\alpha + \beta)$ on (most of) $\mathbb{A}^1$ (undefined where $\alpha = -\beta$).

According to Voskresenskiĭ [39], when $V = \mathbb{G}_m$ one should (at least generically) expect a degree 1 map for each positive integer n, i.e., a birational isomorphism between $(\mathbb{G}_m)_n$ and $\mathbb{A}^{\varphi(n)}$ defined over k. (Conjectures that certain symmetric functions should give dominant maps of low degree were given in [1], but counterexamples to these conjectures were then given in [27].) The next interesting case (because $n/\varphi(n)$, which measures the security per bit, is larger than for the case n=6) is when n=30. The rationality of the algebraic torus $(\mathbb{G}_m)_{30}$ is an open question (in characteristic zero, and over finite fields, for example).

Another question that deserves more research is the security of cryptosystems based on primitive subgroups V_n, or the essentially equivalent question of the security of discrete log cryptography over extension fields $\mathbb{F}_{q^n}$ when n>1. In the abelian variety (or elliptic curve) case this is studied in [11], while the $\mathbb{G}_m$ case is studied in [14]. In addition, Joux et al. [17] and [18] recently obtained variants of the function field and number field sieve that have implications for the security of the discrete log problem for abelian varieties in low characteristic, and for $(\mathbb{G}_m)_{30}$ over $\mathbb{F}_q$ when, for example, q is a 32-bit prime. We encourage further study of the security of cryptosystems based on primitive subgroups.

We also raise the question of finding other applications for varieties $\mathcal{I} \otimes_{\mathcal{O}} V$, in cryptography or elsewhere, including considering other group varieties V and/or other G_k-modules $\mathcal{I}$.

ACKNOWLEDGMENTS

The author was supported by NSA grant H98230-07-1-0039. I thank Karl Rubin for help with the paper, and one of the referees for helpful comments.

REFERENCES

1. W. Bosma, J. Hutton, E.R. Verheul, Looking beyond XTR, in: Advances in Cryptology — Asiacrypt 2002, in: Lect. Notes in Comp. Sci., vol. 2501, Springer, Berlin, 2002, pp. 46–63.
2. A.E. Brouwer, R. Pellikaan, E.R. Verheul, Doing more with fewer bits, in: Advances in Cryptology — Asiacrypt '99, in: Lect. Notes in Comp. Sci., vol. 1716, Springer, Berlin, 1999, pp. 321–332.
3. B. Conrad, Gross–Zagier revisited, in: Heegner points and Rankin L-series, in: Math. Sci. Res. Inst. Pub., vol. 49, Cambridge Univ. Press, Cambridge, 2004, pp. 67–163.
4. I. Dech´ene, Generalized Jacobians in cryptography, Ph.D. Thesis, McGill University, 2005. ` http://www.math.uwaterloo.ca/~idechene/ Dechene thesis. pdf.
5. C. Diem, A study on theoretical and practical aspects of Weil-restrictions of varieties, Dissertation, 2001. http://www.math.uni-leipzig.de/ ~diem/dissertation diem.dvi.
6. C. Diem, N. Naumann, on the structure of Weil restrictions of abelian varieties, J. Ramanujan Math. Soc. 18 (2003) 153–174.
7. G. Frey, How to disguise an elliptic curve (Weil descent), lecture at ECC '98. http://www.cacr.math.uwaterloo.ca/conferences/1998/ecc98/ frey.ps.
8. G. Frey, Applications of arithmetical geometry to cryptographic constructions, in: D. Jungnickel, H. Niederreiter (Eds.), Finite fields and applications, Augsburg, 1999, Springer, Berlin, 2001, and pp. 128–161.

9. S.D. Galbraith, F. Hess, N.P. Smart, Extending the GHS Weil descent attack, in: L. Knudsen (Ed.), Advances in Cryptology — EUROCRYPT 2002, in: Lect. Notes in Comp. Sci., vol. 2332, Springer, Berlin, 2002, pp. 29–44.
10. S.D. Galbraith, B.A. Smith, Discrete logarithms in generalized Jacobians, preprint. http://arxiv.org/abs/math/0610073.
11. P. Gaudry, Index calculus for abelian varieties and the elliptic curve discrete logarithm problem, preprint, October 26, 2004 version.
12. P. Gaudry, F. Hess, N.P. Smart, Constructive and destructive facets of Weil descent on elliptic curves, J. Cryptology 15 (2002) 19–46.
13. G. Gong, L. Harn, Public-key cryptosystems based on cubic finite field extensions, IEEE Trans. Inform. Theory 45 (1999) 2601–2605.
14. R. Granger, F. Vercauteren, On the discrete logarithm problem on algebraic tori, in: V. Shoup (Ed.), Advances in Cryptology — CRYPTO 2005, in: Lect. Notes in Comp. Sci., vol. 3621, Springer, Berlin, 2005, pp. 66–85.
15. A. Grothendieck, J.-L. Verdier, expose IV of Th ́eorie des topos et cohomologie ́ etale des sch ́emas. Tome 1: Th ́eorie des topos, in: M. Artin, ́A. Grothendieck, J.-L. Verdier (Eds.), Seminaire de G ́eom ́etrie Alg ́ebrique du Bois-Marie 1963–1964 (SGA 4), in: Lecture Notes in Math., ́vol. 269, Springer, Berlin, New York, 1972.
16. E. Howe, Isogeny classes of abelian varieties with no principal polarizations, in: G. van der Geer, C. Faber, F. Oort (Eds.), Moduli of Abelian Varieties, Texel Island, 1999, in: Progress in Math., vol. 195, Birkhauser, Basel, 2001, pp. 203–216.
17. A. Joux, R. Lercier, The Function Field Sieve in the Medium Prime Case, in: S. Vaudenay (Ed.), Advances in Cryptology — Eurocrypt 2006, in: Lect. Notes in Comp. Sci., vol. 4004, Springer, Berlin, 2006, pp. 254–270.
18. A. Joux, R. Lercier, N. Smart, F. Vercauteren, The number field sieve in the medium prime case, in: C. Dwork (Ed.), Advances in Cryptology — CRYPTO 2006, in: Lect. Notes in Comp. Sci., vol. 4117, Springer, Berlin, 2006, pp. 323–341.
19. T. Lange, Trace zero subvarieties of genus 2 curves for cryptosystems, J. Ramanujan Math. Soc. 19 (2004) 1–19.
20. A.K. Lenstra, E.R. Verheul, The XTR public key system, in: M. Bellare (Ed.), Advances in Cryptology — CRYPTO 2000, in: Lect. Notes in Comp. Sci., vol. 1880, Springer, Berlin, 2000, pp. 1–19.
21. J.S. Milne, on the arithmetic of abelian varieties, Invent. Math. 17 (1972) 177–190.
22. B. Mazur, K. Rubin, Finding large Selmer rank via an arithmetic theory of local constants, Ann. of Math. 166 (2007) 579–612.
23. B. Mazur, K. Rubin, A. Silverberg, Twisting commutative algebraic groups, J. Algebra 314 (2007) 419–438.
24. W.B. Muller, W. N ̈obauer, some remarks on public-key cryptosystems, Studia Sci. Math. Hungar. 16 (1981) 71–76.
25. N. Naumann, Weil–Restriktion abelscher Varietaten, Diplomarbeit, Universit ̈at Essen, 1999.

26. K. Rubin, A. Silverberg, Supersingular abelian varieties in cryptology, in: M. Yung (Ed.), Advances in Cryptology — CRYPTO 2002, in: Lect. Notes in Comp. Sci., vol. 2442, Springer, Berlin, 2002, pp. 336–353.

27. K. Rubin, A. Silverberg, Torus-based cryptography, in: D. Boneh (Ed.), Advances in Cryptology — CRYPTO 2003, in: Lect. Notes in Comp. Sci., vol. 2729, Springer, Berlin, 2003, pp. 349–365.

28. K. Rubin, A. Silverberg, Algebraic tori in cryptography, in: A. van der Poorten, A. Stein (Eds.), High Primes and Misdemeanours: Lectures in Honour of the 60th Birthday of Hugh Cowie Williams, in: Fields Institute Communications Series, vol. 41, AMS, Providence, RI, 2004, pp. 317–326.

29. K. Rubin, A. Silverberg, Using primitive subgroups to do more with fewer bits, in: D. Buell (Ed.), Proceedings of Algorithmic Number Theory, 6th International Symposium, ANTS-VI, in: Lect. Notes in Comp. Sci., vol. 3076, Springer, Berlin, 2004, pp. 18–41.

30. K. Rubin, A. Silverberg, Compression in finite fields and torus-based cryptography, SIAM J. Comput. 37 (2008) 1401–1428.

31. J.-P. Serre, Complex multiplication, in: J.W.S. Cassels, A. Frohlich (Eds.), Algebraic Number Theory, Thompson Book Co., Washington, DC, ¨ 1967, pp. 292–296.

32. J.-P. Serre, Linear representations of finite groups, in: Graduate Texts in Mathematics, vol. 42, Springer, New York, 1977.

33. J.-P. Serre, Cohomologie galoisienne, cinquieme` edition, r´evis´ee et compl´et´ee, in: Lecture Notes in Math., vol. 5, Springer, Berlin, 1994.

34. A. Silverberg, Compression for trace zero subgroups of elliptic curves, Trends in Mathematics 8 (2005) 93–100.

35. J.H. Silverman, The arithmetic of elliptic curves, in: Graduate Texts in Mathematics, vol. 106, Springer, New York, 1986.

36. P.J. Smith, M.J.J. Lennon, LUC: A new public key system, in: E.G. Dougall (Ed.), Proceedings of the IFIP TC11 Ninth International Conference on Information Security IFIP/Sec '93, North-Holland, Amsterdam, 1993, pp. 103–117.

37. P. Smith, C. Skinner, A public-key cryptosystem and a digital signature system based on the Lucas function analogue to discrete logarithms, in: J. Pieprzyk, R. Safavi-Naini (Eds.), Advances in Cryptology — Asiacrypt 1994, in: Lect. Notes in Comp. Sci., vol. 917, Springer, Berlin, 1995, pp. 357–364.

38. W.A. Stein, Shafarevich-Tate groups of nonsquare order, in: J. Cremona, J-C. Lario, J. Quer, K. Ribet (Eds.), Modular Curves and Abelian Varieties, in: Progr. Math., vol. 224, Birkhauser, Basel, 2004, pp. 277–289.

39. V.E. Voskresenski˘ı, Algebraic groups and their birational invariants, in: Translations of Mathematical Monographs, vol. 179, AMS, Providence, RI, 1998.

40. A. Weil, Adeles and algebraic groups, in: Progress in Math., vol. 23, Birkhauser, Boston, 1982.

41. H.C. Williams, A p + 1 method of factoring, Math. Comp. 39 (1982) 225–234.

42. H.C. Williams, Some public-key crypto-functions as intractable as factorization, Cryptologia 9 (1985) 223–237.

CITATION

1. A. Silverberg, Applications to cryptography of twisting commutative algebraic groups, Discrete Applied Mathematics, Volume 156, Issue 16, 6 September 2008, Pages 3122-3138, ISSN 0166-218X, http://dx.doi.org/10.1016/j.dam.2008.01.024.

Image Mathematics—Mathematical Intervening Principle Based on "Yin Yang Wu Xing" Theory in Traditional Chinese Mathematics (I)

Yingshan Zhang[1] and Weilan Shao[2]

[1]School of Finance and Statistics, East China Normal University, Shanghai, China
[2]The College English Teaching and Researching Department, Xinxiang University, Xinxiang, China

4

ABSTRACT

By using mathematical reasoning, this paper demonstrates the mathematical intervening principle: "Virtual disease is to fill his mother but real disease is to rush down his son" (虚则补其母, 实则泄其子) and "Strong inhibition of the same time, support the weak" (抑强扶弱) based on "Yin Yang Wu Xing" Theory in image mathematics of Traditional Chinese Mathematics (TCMath). We defined generalized relations and generalized reasoning, introduced the concept of steady multilateral systems with two non-compatibility relations, and discussed its energy properties. Later based on the intervention principle in image mathematics of TCMath and treated the research object of the image mathematics as a steady multilateral system, it has been proved that the mathematical intervening principle is true. The kernel of this paper is the existence and reasoning of the non-compatibility relations in steady multilateral systems, and it accords with the oriental thinking model.

MAIN DIFFERENCES BETWEEN TRADITIONAL CHINESE MATHEMATICS AND WESTERN MATHEMATICS

In Western Mathematics (mathematics; Greek: $\mu \alpha \theta \eta \mu \alpha \tau \iota \kappa$), the word comes from the ancient Greek in the west of the $\mu \theta \eta \mu \alpha$ (mathēma), it's have learning, studying, science, and another relatively narrow meaning and technical sense-"mathematics study", even in its neck comes in. The adjective $\mu \alpha \theta \eta \mu \alpha \tau \iota \kappa$ (mathē- matikos), meaning and learning about or for the hard, also can be used to index of learning. In English on the function of the plural form of, and in French surface mathematiques plural form, can be back to Latin neutral plural mathematica, due to the Greek plural $\tau \alpha \mu \alpha \theta \eta \mu \alpha \tau \iota$ κ (tamathēmatika), the Greek by Aristotle brought refers to "all things several" concept. (Latin: Mathemetica) original intention is number and count of the technology.

According to the understanding of the now in Western mathematics, mathematics is the real world number relationship and the form of the space science. Say simply, is the study of the form and number of science. The start of the study is always from the Axiomatic system. Any Axiomatic system of form and number comes from the observations. Because of the demand on life and labor, and even the most original nationality, also know simple count, and the fingers or physical count development to use digital count. Basic mathematics knowledge and use always individual and community essential to life. The basic concept of the refining is as early as in ancient Egypt, Mesopotamia, and in ancient India of ancient mathematical text and see in considerable. Since then, its development will continue to have a modest progress, and the 16th century until the Renaissance, because of the new scientific and found interactions and the generated mathematical innovation leads to the knowledge of the acceleration, until today.

Mathematics is used in different areas of the world, including science, engineering, medicine and economics, etc. The application of mathematics in these fields are usually called applied mathematics, sometimes also stir up new mathematical discovery, and led to the development of new subject. Mathematician also study didn't any application value of pure mathematics, even if the application is found in often after.

The Boolean school, founded in the 1930s in France by Boolean, thinks: mathematics, at least pure mathematics, is the study of the theory of abstract structure. Structure, it is the initial concept and the deduction system of Axiom. Boolean school also thinks that, there are three basic kinds of abstract structure: algebraic structure (group, ring, the domain...), sequence structure (partial order, all the sequence...), the topological structure (neighborhood, limit, connectivity, dimension...). Mathematics is a kind of transformation, an abstract model, and a sign system, the real world converted into mathematical model, using mathematical language describe them later, after operation, the results can back, explained in the real world specific scientific.

There is a basic logic that the human is everything, they can be observed to establish an Axiom system, make nature operations according to the Axiom system by human assumption. But in the traditional Chinese philosophy, always think that humans are the small, they cannot establish a set of rules, let nature operation in accordance with the running rules of the human assumption. In front to nature, human's only doing things is its behavior request, with its development with nature.

In other words, the western mathematics thinks the mathematical object of study is a simple system, which can be observed to establish an Axiom system, and then logical analysis. It is because a simple system can be assumed. But the image mathematics of the traditional Chinese mathematics thinks the mathematical object of study is a complex system, human can't do specific research object hypothesis, humans can only be clear, for general object of study (model-free), what kind of logic structure analysis can reach the humans to understand the research object of certain relations. It is because a complex system cannot be assumed.

Simple said: the western mathematics deals directly with the research object through the directly observed, but the image mathematics of the traditional Chinese mathematics researches object through the relationship between indirect processing analyses.

In fact, Western mathematics late nineteenth century was introduced into China, initially, "mathematics" to be directly translated as "arithmetic (算术)", and then said "arithmetic learn (算学)", and then they changed to "mathematics (数学)" words. But the ancient Chinese for this concept, in 3000 years ago, has been officially use the word "Gua (卦)" as the form, the "Xiang (象)" as the number. In the Yi-Jing ("易经"), this "mathematics (数学)" concept is defined as "Image Mathematics (象数学)". It is part of the Traditional Chinese Mathematics (TCMath). This article mainly concerns image mathematical content in TCMath, so also said the image mathematics as TCMath. Image mathematics generally contains hexagrams ("Gua (卦)") and images ("Xiang (象)") two content. The hexagrams ("Gua (卦)") is only the hexagrams mathematical symbols, which there is not the size since the size of number is about the definition of human beings. In general, the research object in the complex system is independent of human definition. The image ("Xiang (象)") is the study way or the calculation method for some mathematical indexes in order to know the objective existence of the fixed a state. The way or the calculation method for some mathematical indexes is independent of the complex system and only is Human's some methods of operation in order to study the relationship of the complex system.

The ancients speak of "mathematics" in Chinese is a word as the way for running (intervening and controlling) a complex system through the analysis of the use of hexagrams ("Gua (卦)") and images ("Xiang (象)"). This and what we now understand the mathematical completely different things. In other words, in the image mathematics of Traditional Chinese Mathematics (TCMath), both intervening and controlling of an engineering are believed to as a complex system. It is because to run an engineering is difficult and complex in which there are the loving relation, the killing relation and the equivalent relation among many Axiom systems. The loving and killing relations are non-compatibility relations, which can compose the whole energy of the system greater than or less than the sum of each part energy of the system, respectively, rarely equal conditions. Mathematics means managing or controlling or intervening for the complex system through the analysis of the use of hexagrams ("Gua (卦)") and images ("Xiang (象)"), and so on. Pursue the goal is the harmonious sustainable of the complex system in order to compose the complex system not outward expansion development. Generally speaking, the assumption involved the behavior of people is not needed since the system is complex.

But, in Western mathematics, mathematics means first through the observation to establish one Axiom system, then performing mathematical inference from the Axiom system. Both obtaining and inference of reasoning are believed to as a simple system, because all the conclusions and definition of are compatible with the Axiom system. Compared with Axiom system speaking, there are true and false. Major mathematical analysis method is to judge true or false from simple assumptions or simple model. It is because to obtain the true and false relationship of one Axiom system is easy and simple in which there is only a compatibility relation or a generalized equivalent relation under one Axiom system assumption. The compatibility relation or generalized equivalent relation can compose the whole energy of the system equal to the sum of each part energy of the system. Thus to obtain or to analyze under one Axiom system assumption can compose the simple system outward expansion development. Therefore, pursue the goal is for obtaining or analyzing in order to compose the simple system outward expansion development. Generally speaking, the various hypothetical models involves the behavior of people. This phenomenon in the image mathematics of TCMath is not allowed since a complex system cannot be supposed. Both true and false cannot be judged if the Axiom system has not been assumed.

Western mathematics using simple assumptions or simple models treats directly mathematical complex system from Microscopic point of view, always destroy the original mathematical complex system's balance, and has none beneficial to mathematical complex system's immunity. Western mathematical intervention method can produce imbalance of mathematical complex system, having strong side effects. Excessively using methods of mathematical intervention for a complex system can easily paralysis the mathematical complex system's immunity, which the debate of mathematical schools under different from Axiom systems is a product of Western mathematics since there are a number of Axiom systems in nature which are different to the people of faith. Using the method of mathematical intervention for a complex system too little can easily produce the mathematical intervention resistance problem.

The image mathematics of TCMath studies the world from the macroscopic point of view, and its target is in order to maintain the original balance of mathematical complex system and in order to enhance the mathematiccal complex system's immunity. The image mathematics believes that each mathematical intervention has onethird of badness. She never encourage government to use mathematical intervention in long term. The ideal way is Wu Wei Er Wu Bu Wu (无为而无不为)—by doing nothing, everything is done. The image mathematics has over 5000-year history. It has almost none side effects or mathematical intervention resistance problem.

After long period of practicing, our ancient mathematical scientists use "Yin Yang Wu Xing" Theory extensively in the image mathematics to explain the origin of mathematical complex system, the law of mathematiccal complex system, mathematical changes, mathematical diagnosis, mathematical prevention, mathematical self-protection and mathematical intervening. It has become an important part of the image mathematics. "Yin Yang Wu Xing" Theory has a strong influence to the formation and development of traditional Chinese mathematical theory. As is known to all, China in recent decades, economy and related mathematical work have made great strides in development. Its reason is difficult to say the introduction of western mathematics, the fact that the Chinese traditional culture is in all kinds of mathematical decision plays a role. Her many mathematical intervening methods come from the traditional Chinese medicine since both human body and mathematical research objects of image mathematics are all complex systems. But, many Chinese and foreign schoolars still have some questions on the reasoning of image mathematics, such as Traditional Chinese Medicine which is due to image mathematics. In this article, we will start to the western world for presentation of image mathematics introduced some mathematical and logic analysis concept.

Zhang's theories, multilateral matrix theory [1] and multilateral system theory [2-19], have given a new and strong mathematical reasoning method from macro (Global) analysis to micro (Local) analysis. He and his colleagues have made some mathematical models and methods of reasoning [20-35], which make the mathematical reasoning of image mathematics possible based on "Yin Yang Wu Xing" Theory [36-38]. This paper will use steady multilateral systems to demonstrate the intervening principle of image mathematics: "Real (mathematical) disease is to rush down his son but virtual (mathematical) disease is to fill his mother" and "Strong inhibition of the same time, support the weak".

The article proceeds as follows. Section 2 contains basic concepts and main theorems of steady multilateral systems while the intervening principle of image mathematics is demonstrated in Section 3. Some discussions in image mathematics are given in Section 4 and conclusions are drawn in Section 5.

BASIC CONCEPT OF STEADY MULTILATERAL SYSTEMS

In the real world, we are enlightened from some concepts and phenomena such as "biosphere", "food chain", "ecological balance" etc. With research and practice, by using the theory of multilateral matrices [1] and analyzing the conditions of symmetry [20-24] and orthogonality [25-35] what a stable complex system must satisfy, in particular, with analyzing the basic conditions what a stable working procedure of good product quality must satisfy [10, 29], we are inspired and find some rules and methods, then present the logic model of analyzing stability of complex systems-steady multilateral systems [2-19]. There are a number of essential reasoning methods based on the stable logic analysis model, such as "transition reasoning", "atavism reasoning", "genetic reasoning" etc. We start and still use concepts and notations in papers [3-6].

Generalized Relations and Reasoning

Let V be a non-empty set and $V \times V = \{(x,y): x \in V, y \in V\}$. The non-empty subset $R \subset V \times V$ is called a relation of V. Image mathematics mainly researches general relation rules for general V rather than for special V. So the general V cannot be supposed. We can only do matter is to research the structure of the set of relations $\Re = \{R_0,...,R_{m-1}\}$.

For any relation set $\Re = \{R_0,...,R_{m-1}\}$, we can define both an inverse relationship of $R_i \in \Re$ and a relation multiplication between $R_i \in \Re$ and $R_j \in \Re$ as follows:

$$R_i^{-1} = \{(x,y):(y,x) \in R_i\}$$

and $R_i * R_j = \{(x,y):$ there is at least a $u \in V$ such that $(x,u) \in R_i$ and $(u,y) \in R_j\}$.

The relation $R_i \in \Re$ is called reasonable if $R_i^{-1} \in \Re$. A generalized reasoning of general V is defined as for $R_i * R_j \neq \varnothing$ there is a relation $R_k \in \Re$ such that $R_i * R_j \subset R_k$.

The generalized reasoning satisfies the associative law of reasoning, i.e. $(R_i * R_j) * R_k = R_i * (R_j * R_k)$. This is the basic requirement of reasoning in TCMath. But there are a lot of reasoning forms which do not satisfy the associative law of reasoning in Western Science. For example, in true and false binary of proposition logic, the associative law does not hold on its reasoning because

$$(false * false) * false = true * false = false$$
$$\neq true = false * true = false * (false * false).$$

Equivalence Relations

Let V be a non-empty set and R_0 be its relation. We call it an equivalence relation, denoted by ~, if the following three conditions are all true:

1) Reflexive: $(x,x) \in R_0$ for all $x \in V$, i.e., $x \sim x$;

2) Symmetric: if $(x,y) \in R_0$, then $(y,x) \in R_0$, i.e., if $x \sim y$, then $y \sim x$;

3) Conveyable (Transitivity): if $(x,y) \in R_0$, $(y,z) \in R_0$, then $(x,z) \in R_0$, i.e., if $x \sim y$, $y \sim z$, then $x \sim z$.

Furthermore, the relation R is called a compatibility relation if there is a non-empty subset $R_1 \subset R$ such that R_1 satisfies at least one of the conditions above. And the relation R is called a non-compatibility relation if there doesn't exist any non-empty subset $R_1 \subset R$ such that R_1 satisfies any one of the conditions above. Any one of compatibility relations can be expanded into an equivalent relation to some extent [2].

Western Science only considers the reasoning under one Axiom system such that only compatibility relation reasoning is researched. However there are many Axiom systems in Nature. Traditional Chinese Science mainly researches the reasoning among many Axiom systems in Nature. Of course, she also considers the reasoning under one Axiom system but she only expands the reasoning as the equivalence relation reasoning.

Two Kinds of Opposite Non-compatibility Relations

Equivalence relations, even compatibility relations, cannot portray the structure of the complex systems clearly. In the following, we consider two non-compatibility relations.

In image mathematics, any Axiom system is not considered, but should first consider to use a logic system. Believe that the rules of Heaven and the behavior of Human can follow the same logic system (天人

合一). This logic system is equivalent to a group of computation. The method is to abide by the selected logic system to the research object classification, without considering the specific content of the research object, namely classification taking images (比类取象). Analysis of the relationship between research object, make relationships with computational reasoning comply with the selected logic system operation. And then in considering the research object of the specific content of the conditions, according to the logic of the selected system operation to solve specific problems. In mathematics, the method of classification taking images is explained in the following Definition 2.1.

Definition 2.1: Suppose that there exists a finite group $G^m = \{g_0,...,g_{m-1}\}$ of order m where g_0 is identity. Let V be a none empty set satisfying that $V = V_{g_0} +...+ V_{g_{m-1}}$ where the notation means that $V = V_{g_0} \cup...\cup V_{g_{m-1}}$, $V_{g_i} \cap V_{g_j} = \varnothing, \forall i \neq j$ (the following the same).

In image mathematics, the V_{g_j} is first called a factor image of group element g_j for any j, and $V = V_{g_0} +...+ V_{g_{m-1}}$ is called a factor space (all "Gua (卦)"). We do not consider the factor size (class variables) and only consider it as mathematical symbols ("Gua (卦)"), such as, 0 or 1, because the size is defined by a human behavior for V, but we have no assumption of V.

A mathematical index of the unknown multivariate function $f(x_{g_0},...,x_{g_{m-1}})$, $j = 0,...,m-1 \, \forall x_{g_j} \in V_{g_j}$ is called a function image of V. All mathematical indexes of the unknown multivariate function f compose of the formation of a new set, namely image space

$$F(V) = F_{\omega_0}(V) +\cdots+ F_{\omega_{g-1}}(V)$$

Where $G^g = \{\omega_0,...,\omega_{g-1}\}$ is also a finite group of order g. The $F_{\omega_j}(V)$ is also called an Axiom system for any j if $F_{\omega_j}(V) \neq \varnothing$ because any Axiom system is the assumption of

$$F(V) = F_{\omega_0}(V) + \cdots + F_{\omega_{g-1}}(V)$$

(or, equivalent, V) in which there is only the compatibility relations, i.e., pursuing the same mathematics index $I(F_{\omega_j}(V))$. We do not consider the special multivariate function f (i.e., special function image) and only consider the calculation way of the general mathematical indexes of f from the factor space $V = V_{g_0} + \ldots + V_{g_{m-1}}$ in order to know some causal relations, because we have no assumption of f. But the size of the data image should be considered if we study specific issues by the general rules of data images.

Say simply, a study of the hexagrams ("Gua (卦)") in image mathematics is to learn the generalized properties of the inputs $x_{g_0}, \ldots, x_{g_{m-1}}$ of any multivariate function f for the given factor space $V = V_{g_0} + \ldots + V_{g_{m-1}}$, such as there are non-size, non-order relation, orthogonal relations and symmetrical relations, and so on. A study of the image ("Xiang (象)") in image mathematics is to learn the generalized properties of all outputs f for the image space $F(V) = F_{\omega_0}(V) + \ldots + F_{\omega_{g-1}}(V)$, such as there are size specific meaning, a sequence of relationship, killing relations, loving relations, equivalent relations, and so on.

Without loss of generality, we put the function image space $F(V) = F_{\omega_0}(V) + \ldots + F_{\omega_{g-1}}(V)$ and the factor image space $V = V_{g_0} + \ldots + V_{g_{m-1}}$, still keep for V because of no assumption of V. In order to study the generalized relations and generalized reasoning, image mathematics researches the following relations.

Denoted $V_{g_i} \times V_{g_j} = \{(x,y) : x \in V_{g_i}, y \in V_{g_j}\}$, where the note × is the usual Cartesian product or cross join. Define relations

$$R_{g_r} = \sum_{g \in G^m} V_g \times V_{gg_r}, r = 0, \cdots, m-1,$$

where $R_{g_0} = R_{g_0}^{-1} = R_{g_0^{-1}}$ is called an equivalence relation of V if g_0 is identity; denoted by ~; $R_{g_s} = R_{g_s}^{-1} = R_{g_s^{-1}}$ is called a symmetrical relation of V if $g_s = g_s^{-1}, s \neq 0$; denoted by $\xleftrightarrow{R_s}$ or $\leftrightarrow$; $R_{g_1} = R_{g_1^{-1}}^{-1}$ is called a neighboring relation of V if $g_1 \neq g_1^{-1}$; denoted by $\xrightarrow{R_1}$ or $\rightarrow$; $R_{g_a} = R_{g_a^{-1}}^{-1} \neq R_{g_a^{-1}}, R_{g_1}, R_{g_1^{-1}}$ is called an alternate (or atavism) relation of V if $g_a \neq g_a^{-1}, g_1, g_1^{-1}, a \geq 2$; denoted by $\xRightarrow{R_a}$ or $\Rightarrow$.

In this case, the equivalence relations and symmetrical relations are compatibility relations but both neighboring relations and alternate relations are non-compatibility relations. For the given relation set $\mathfrak{R} = \{R_{g_0}, \ldots, R_{g_{m-1}}\}$, these relations R_{g_i} are all reasoning relations since the relation $R_{g_i}^{-1} = R_{g_i^{-1}} \in \mathfrak{R}$ if $R_{g_i} \in \mathfrak{R}$.

The equivalence relation R_{g_0}, symmetrical relations R_{g_s}, neighboring relation R_{g_1} and alternate relations R_{g_a} are all the possible relations for the method of classification taking images. In this paper, we mainly consider the equivalence relation R_{g_0}, neighboring relation R_{g_1} and alternate relations R_{g_a}.

There is an unique generalized reasoning between the two kinds of opposite non-compatibility relations for case m = 5. For example, let V be a none empty set, there are two kinds of opposite relations: the neighboring relation R_1, denoted $\rightarrow$ and the alternate (or atavism) relation R_2, denoted $\Rightarrow$. The logic reasoning architecture [2-19] of "Yin Yang Wu Xing" Theory in Ancient China is equivalent to the following reasoning:

1. If $x \rightarrow y, y \rightarrow z$, then $x \Rightarrow z$; i.e., if $(x,y) \in R_1$, $(y,z) \in R_1$, then $(x,z) \in R_2$; or, $R_1 * R_1 \subset R_2$; $\Leftrightarrow$ if $x \rightarrow y, x \Rightarrow z$, then $y \rightarrow z$; i.e., if

$(x, y) \in R_1$, $(x, z) \in R_2$, then $(y, z) \in R_1$; or, $R_1^{-1} * R_2 \subset R_1$; $\Leftrightarrow$ if $x \Rightarrow z$, $y \to z$, then $x \to y$; i.e., if $(x, z) \in R_2$, $(y, z) \in R_1$, then $(x, y) \in R_1$; or, $R_2 * R_1^{-1} \subset R_1$.

2. If $x \Rightarrow y$, $y \Rightarrow z$ then $z \to x$; i.e., if $(x, y) \in R_2$, $(y, z) \in R_2$, then $(z, x) \in R_1$; or, $R_2 * R_2 \subset R_1^{-1}$; $\Leftrightarrow$ if $z \to x$, $x \Rightarrow y$, then $y \Rightarrow z$; i.e., if $(z, x) \in R_1$, $(x, y) \in R_2$ then $(y, z) \in R_2$; or, $R_1 * R_2 \subset R_2^{-1}$; $\Leftrightarrow$ if $y \Rightarrow z$, $z \to x$ then $y \Rightarrow z$; i.e., if $(y, z) \in R_2$, $(z, x) \in R_1$, then $(y, z) \in R_2$; or, $R_2 * R_1 \subset R_2^{-1}$.

Let $R_3 = R_2^{-1}$ and $R_4 = R_1^{-1}$. Then above reasoning is equivalent to the calculating as follows:

$$R_i * R_j \subset R_{\mathrm{mod}(i+j,5)}, \forall i, j \in \{1, 2, 3, 4\}$$

Where the $\mathrm{mod}(i+j, 5)$ is the addition of module 5.

Two kinds of opposite relations cannot be exist separately. Such reasoning can be expressed in Figure 1. The first triangle reasoning is known as a jumping-transition reasoning, while the second triangle reasoning is known as an atavism reasoning. Reasoning method is a triangle on both sides decided to any third side. Both neighboring relations and alternate relations are not compatibility relations, of course, none equivalence relations, called non-compatibility relations.

Genetic Reasoning

Let V be a none empty set with the equivalent relation R_0, the neighboring relation R_1 and the alternate relations $R_a \neq R_1^{-1}$, $a > 1$. Then a genetic reasoning is defined as follows:

1. If $x \sim y$, $y \to z$, then $x \to z$; i.e., if $(x,y) \in R_0$, $(y,z) \in R_1$, then $(x,z) \in R_1$; or, $R_0 * R_1 \subset R_1$;

2. If $x \sim y$, $y \Rightarrow z$, then $x \Rightarrow z$; i.e., if $(x,y) \in R_0$, $(y,z) \in R_2$, then $(x,z) \in R_2$; or, $R_0 * R_2 \subset R_2$;

3. If $x \to y$, $y \sim z$, then $x \to z$; i.e., if $(x,y) \in R_1$, $(y,z) \in R_0$, then $(x,z) \in R_1$; or, $R_1 * R_0 \subset R_1$;

4. If $x \Rightarrow y$, $y \sim z$, then $x \Rightarrow z$; i.e., if $(x,y) \in R_2$, $(y,z) \in R_0$, then $(x,z) \in R_2$; or, $R_2 * R_0 \subset R_2$.

The genetic reasoning is equivalent to that

$$R_0 * R_j = R_j * R_0 = R_j, \forall j \in \{0,1,\cdots,m-1\} = G_0^m$$

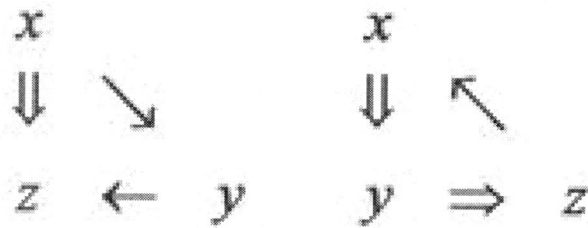

Figure 1: Triangle reasoning.

Since $R_0 * R_j \supset R_j$, $R_j * R_0 \supset R_j$. The genetic reasoning is equivalent to that there is a group $G_0^m = \{0,...,m-1\}$ with the operation $*$ such that V can be cut into $V = V_0 + ... + V_{m-1}$ where V_i may be an empty set and the corresponding relations of reasoning can be written as the forms as follows

$$R_r = \sum_{i=0}^{m-1} V_i \times V_{i*r}, \, r = 0, \cdots, m-1,$$

Satisfying

$$R_i * R_j \subset R_{i*j}, \, \forall i, j \in G_0^m.$$

Steady Multilateral Systems

For a none empty set V and it's a relation set $\Re = \{R_0, ..., R_{m-1}\}$, the form $(V, \Re)$ (or simply, V) is called a multilateral system [2-19], if $(V, \Re)$ satisfies the following properties:

1. $R_0 + ... + R_{m-1} \subset V \times V$, i.e. $R_i \cap R_j = \emptyset, \forall i \neq j$.

2. $R_0 * R_j = R_j * R_0 = R_j, \, \forall j \in \{0, 1, \cdots, m-1\} = G_0^m$

3. The relation $R_i^{-1} \in \Re$ if $R_i \in \Re$.

4. For $R_i * R_j \neq \emptyset$, there is a relation $R_k \in \Re$ such that $R_i * R_j \subset R_k$.

The 4) is called the reasoning, the 1) the uniqueness of reasoning, the 2) the hereditary of reasoning (or genetic reasoning) and the 3) the equivalent property of reasoning of both relations $R_i \in \Re$ and $R_i^{-1} \in \Re$, i.e., the reasoning of $R_i \in \Re$ is equivalent to the reasoning of $R_i^{-1} \in \Re$. In this case, the two-relation set $\{R_i, R_i^{-1}\}$ is a lateral relation of $(V, \Re)$. The R_0 is called an equivalence relation. The multilateral system $(V, \Re)$ can be written as $(V_0 + ... + V_{n-1}, \{R_0 ..., R_{m-1}\})$. Furthermore, the V and $\Re$ are called the state space and relation set considered of $(V, \Re)$, respectively. For a multilateral system $(V, \Re)$, it is called complete (or, perfect) if " $\subset$ " changes into "=". And it is called complex if there exists at least a non-compatibility relation $R_i \in \Re$. In this case, the multilateral system is also called a logic analysis model of complex systems.

Let R_1 be a non-compatibility relation. A complex multilateral system

$$(V, \Re) = \left(V_0 + \cdots + V_{n-1}, \{R_0, \cdots, R_{m-1}\} \right)$$

Is said as a steady multilateral system (or, a stable multilateral system) if there exists a number n such that $R_1^{*n} = R_0$ where $R_1^{*n} = R_1 \overset{n}{*} \ast \ast R_1$. The condition is equivalent to there is a chain $x_1, \ldots, x_n \in V$ such that $(x_1, x_2) \in R_1, \ldots, (x_n, x_1) \in R_1$, or $x_1 \to x_2 \to \ldots \to x_n \to x_1$. The steady multilateral system is equivalent to the complete multilateral system. The stability definition given above, for a relatively stable system, is most essential. If there is not the chain or circle, then there will be some elements without causes or some elements without results in a system. Thus, this system is to be in the state of finding its results or causes, i.e., this system will fall into an unstable state, and there is not any stability to say.

Theorem 2.1: The system $(V, \Re)$ is a multilateral system if and only if there exists a finite group

$G^m = \{g_0, \ldots, g_{m-1}\}$ of order m where g_0 is identity such that the relation set $\Re = \{R_{g_0}, \ldots, R_{g_{m-1}}\}$ satisfying $R_{gi} \ast R_{gj} \subset R_{gigj}, \forall i, j \in \{0, 1, \cdots, m-1\}$ In this case, the multilateral system $(V, \Re)$ can be written as $(V_{g_0} + \ldots + V_{g_{m-1}}, \{R_{g_0}, \ldots, R_{g_{m-1}}\})$ satisfying $R_{gr} = \sum_{g \in G^m} V_g \times V_{ggr}, r = 0, \cdots, m-1$ Where V_{gi} may be an empty set.

Theorem 2.2: If the multilateral system $(V, \Re) = (V_0 + \ldots + V_{n-1}, \{R_0, \ldots, R_{m-1}\})$ is a steady multilateral system, then n=m and $\Re = \{R_0, \ldots, R_{m-1}\}$ is a finite group of order m about the relation multiplication $R_i \ast R_j = R_k$ where V_i must be a non-empty set.

Definition 2.2: Suppose that a multilateral system $(V, \mathfrak{R})$ can be written as $(V_{g_0} + \ldots + V_{g_{m-1}}, \{R_{g_0}, \ldots, R_{g_{m-1}}\})$ satisfying

$$R_{g_r} = \sum_{g \in G^m} V_g \times V_{gg_r}, \quad r = 0, \cdots, m-1$$

and

$$R_{g_i} * R_{g_j} \subset R_{g_i g_j}, \quad \forall i, j \in \{0, 1, \cdots, m-1\}.$$

The group $G^m = \{g_0, \ldots, g_{m-1}\}$ of order m where g_0 is identity is called the representation group of the multilateral system $(V, \mathfrak{R})$. The representing function of R_{g_r} is defined as follows

$$I\left(R_{g_r}\right) = \left\{(x, y) : x^{-1} y = g_r, x, y \in G^m\right\}, r = 0, \cdots, m-1.$$

Let multilateral systems $(V^i, \mathfrak{R}^i), i = 1, 2$ be with two representation groups $G_{i'}$, i=1, 2, respectively. Both multilateral systems $(V^i, \mathfrak{R}^i), i = 1, 2$ are called isomorphic if the two representation groups $G_{i'}$, i=1, 2 are isomorphic.

Theorems 2.1 and 2.2 and Definitions 2.1 and 2.2 are key concepts in multilateral system theory because they show the classification taking images as the basic method. In the following, introduce two basic models to illustrate the method.

Theorem 2.3: Suppose that $G_0^2 = \{0, 1\}$ with multiplication table i.e., the multiplication of G_0^2 is the addition of module 2. In other words, $i * j = \mod(i + j, 2)$.

*	0	1
0	0	1
1	1	0

And assume that $(V, \Re) = (V_0 + V_1, \{R_0, R_1\})$ satisfying

$$R_r = \sum_{i=0}^{1} V_i \times V_{\mathrm{mod}(i+r,2)}, \quad r = 0, 1,$$

$$R_i * R_j \subset R_{\mathrm{mod}(i+j,2)}, \quad \forall i, j \in G_0^2.$$

Then $(V, \Re)$ is a steady multilateral system with one equivalent relation R_0 and one symmetrical relation R_1 which is a simple system since there is not any noncompatibility relation. In other words, the relations R_i's are the simple forms as follows:

$$I(R_0) = \{(0,0),(1,1)\}, \quad I(R_1) = \{(0,1),(1,0)\}$$

Where (i, j) is corresponding to $V_i \times V_j$.

It will be proved that the steady multilateral system in Theorem 2.3 is the reasoning model of "Tao" (道) in TCMath if there are two energy functions $\varphi(V_0)$ and $\varphi(V_1)$ satisfying $\varphi(V_0) > \varphi(V_1)$, called Dao model, denoted by V^2.

Theorem 2.4: For each element x in a steady multilateral system V with two non-compatibility relations, there exist five equivalence classes below:

$$X = \{y \in V \mid y \sim x\}, \ X_S = \{y \in V \mid x \rightarrow y\},$$

$$X_K = \{y \in V \mid x \Rightarrow y\}, \ K_X = \{y \in V \mid y \Rightarrow x\},$$

$$S_X = \{y \in V \mid y \rightarrow x\}$$

Which the five equivalence classes have relations in Figure 2.

It can be proved that the steady multilateral system in Theorem 2.4 is the reasoning model of "Yin Yang Wu Xing" in TCMath if there are five energy functions (Defined in Section 3) $\varphi(K_X)\varphi(X_K), \varphi(X_S), \varphi(X)$, and $\varphi(S_X)$ satisfying $\varphi(X_K) > \varphi(X_S) > \varphi(X) > \varphi(K_X) > \varphi(S_X)$ called Wu-Xing model, denoted by V_5.

By Definition 2.2, the Wu-Xing model V^5 can be written as follows:

Define $V_0 = X$, $V_1 = X_S$, $V_2 = X_K$, $V_3 = K_X$, $V_4 = S_X$, corresponding to wood, fire, soil, metal, water, and assume $V = V_0 + V_1 + ... + V_4$ and $\mathfrak{R} = \{R_0, R_1, ..., R_4\}$, satisfying

$$R_r = \sum_{i=0}^{4} V_i \times V_{\mathrm{mod}(i+r,5)}, \, r \in G_0^5,$$

$$R_i * R_j \subset R_{\mathrm{mod}(i+j,5)}, \, \forall i, j \in G_0^5$$

i.e., the relation multiplication of V^5 is isomorphic to the addition of module 5. Then V^5 is a steady multilateral system with one equivalent relation R_0 and two non-compatibility relations $R_1 = R_4^{-1}$ and $R_2 = R_3^{-1}$.

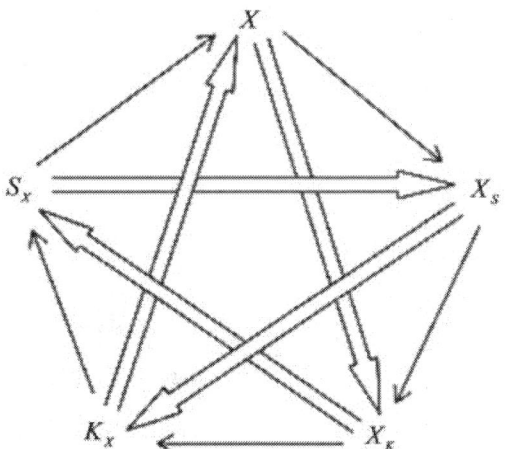

Figure 2: The method of finding Wu-Xing.

These Theorems can been found in [1-6, 11-16]. Figure in Theorem 2.4 is the Figure of "Yin Yang Wu Xing" Theory in Ancient China. The steady multilateral system V with two non-compatibility relations is equivalent to the logic architecture of reasoning model of "Yin Yang Wu Xing" Theory in Ancient China. What describes the general method of complex systems can be used in mathematical complex system.

RELATIONSHIP ANALYSIS OF STEADY MULTILATERAL SYSTEMS

Energy of a Multilateral System

Energy concept is an important concept in Physics. Now, we introduce this concept to the multilateral systems (or image mathematics) and use these concepts to deal with the multilateral system diseases (mathematical index too bad or too good).In mathematics, a multilateral system is said to have Energy (or Dynamic) if there is a none negative function $\varphi(*)$ which makes every subsystem meaningful of the multilateral system.

For two subsystems V_i and V_j of multilateral system V, denote $V_i \rightarrow V_j$ (or $V_i \Rightarrow V_j$, or $V_i \sim V_j$, or $V_i \leftrightarrow V_j$) means $x_i \rightarrow x_j$, $\forall x_i \in V_i$, $x_j \in x_j$ (or $x_i \Rightarrow x_j$, $\forall x_i \in V_i$, $x_j \in V_j$, or $x_i \sim x_j$, $\forall x_i \in V_i$, $x_j \in V_j$, or $x_i \leftrightarrow x_j$, $\forall x_i \in V_i$, $x_j \in V_j$).

For subsystems V_i and V_j where $V_i \cap V_j = \varnothing$, $\forall i \neq j$. Let $\varphi(V_i)$, $\varphi(V_j)$ and $\varphi(V_i, V_j)$ be the energy function of V_i, the energy function of V_j and the total energy of both V_i and V_j, respectively.

For an equivalence relation $V_i \sim V_j$, if $\varphi(V_i, V_j) = \varphi(V_i) + \varphi(V_j)$ (the normal state of the energy of $V_i \sim V_j$), then the neighboring relation $V_i \sim V_j$ is called that V_i likes V_j which means that V_i is similar to V_j. In this case, the V_i is also called the brother of V_j while the V_j is also called the brother of V_i. In the causal model, the V_i is called the similar family member of V_j while the V_j is also called the similar family member of V_i. There are not any causal relation considered between V_i and V_j.

For a symmetrical relation $V_i \leftrightarrow V_j$, if $\varphi(V_i, V_j) = \varphi(V_j, V_i) = \varphi(V_i) + \varphi(V_j) + \tau(V_i, V_j)$,

(The normal state of the energy of $V_i \leftrightarrow V_j$) where the $\tau(V_i, V_j)$ is an interaction of V_i and V_j satisfying $\tau(V_i, V_j) = \tau(V_j, V_i)$, then the symmetrical relation $V_i \leftrightarrow V_j$ is called that V_i is corresponding to V_j which means that V_i is positively (or non-negatively) corresponding to V_j if $\tau(V_i, V_j) > 0$ (or $\tau(V_i, V_j) \geq 0$) and that V_i is negatively corresponding to V_j if $\tau(V_i, V_j) < 0$. In this case, the V_i is also called the counterpart of V_j while the V_j is also called the counterpart of V_i. In the causal model, the V_i is called the reciprocal causation of V_j while the V_j is also called the reciprocal causation of V_i. There is a reciprocal causation relation considered between V_i and V_j.

For an neighboring relation $V_i \rightarrow V_j$, if

$$\varphi(V_i, V_j) > \varphi(V_j, V_i) > \varphi(V_i) + \varphi(V_j),$$

(The normal state of the energy of $V_i \rightarrow V_j$), then the neighboring relation $V_i \rightarrow V_j$ is called that V_i bears (or loves) V_j [or that V_j is born by (or is loved by) V_i] which means that V_i is beneficial on V_i each other. In this case, the V_i is called the mother of V_j while the V_j is called the son of V_i. In the causal model, the V_i is called the beneficial cause of V_j while the V_j is called the beneficial effect of V_i.

For an alternate relation $V_i \Rightarrow V_j$, if

$$\varphi(V_i, V_j) < \varphi(V_j, V_i) < \varphi(V_i) + \varphi(V_j),$$

(The normal state of the energy of $V_i \Rightarrow V_j$), then the alternate relation $V_i \Rightarrow V_j$ is called as that V_i kills (or hates) V_j [or that V_j is killed by (or is hated by) V_i] which means that V_i is harmful on V_j each other. In this case, the V_i is called the bane of V_j while the V_j is called the prisoner of V_i. In the causal model, the V_i is called the harmful cause of V_j while the V_j is called the harmful effect of V_i.

In the future, if not otherwise stated, any equivalence relation is the liking relation, any symmetrical relation is the reciprocal causation relation, any neighboring relation is the born relation (or the loving relation), and any alternate relation is the killing relation.

Suppose V is a steady multilateral system having energy, then V in the multilateral system during normal operation, its energy function for any subsystem of the multilateral system has an average (or expected value in Statistics), this state is called normal when the energy function is nearly to the average. Normal state is the better state.

That a subsystem of a multilateral system is not running properly (or, abnormal), is that the energy deviation from the average of the subsystems is too large (or too big), the high (mathematical real disease or economic overheating or real disease) or the low (mathematical virtual disease or economic downturn or virtual disease). Both mathematical real disease and mathematical virtual disease are all diseases of mathematical complex systems.

In a subsystem of a multilateral system being not running properly, if this sub-system through the energy of external forces increase or decrease, making them return to the average (or expected value), this method is called intervention (or making a mathematical treatment) to the multilateral system.

The purpose of intervention is to make the multilateral system return to normal state. The method of intervention is to increase or decrease the energy of a subsystem.

What kind of intervening should follow the principle to treat it? For example, Western economics emphasizes direct intervening, but the indirect intervening of oriental economics is required. In mathematics, which is more reasonable?

Based on this idea, many issues are worth further discussion. For example, if an economic intervening has been done to an economic society, what situation will happen?

Intervention Rule of a Multilateral System

For a steady multilateral system V with two non-compatibility relations, suppose that there is an external force (or an intervening force) on the subsystem X of V which makes it the energy $\varphi(X)$ change increment $\Delta\varphi(X)$, then the energies $\varphi(X_S)$, $\varphi(X_K)$, $\varphi(K_X)$ $\varphi(S_X)$ of other subsystems X_S, X_K, K_X, S_X (defined in Theorem 2.4) of V will be changed by the increments $\Delta\varphi(X_S)$, $\Delta\varphi(X_K)$, $\Delta\varphi(K_X)$ and $\Delta\varphi(S_X)$, respectively.

It is said that the multilateral system has the capability of intervention reaction if the multilateral system has capability to response the intervention force.

If a subsystem X of multilateral system is intervened, then the energies of the subsystems X_S and S_X which have neighboring relations to X will change in the same direction of the force outside on X. We call them beneficiaries. But the energies of the subsystems X_K and K_X which have alternate relations to X will change in the opposite direction of the force outside on X. We call them victims.

In general, there is an essential principle of intervenetion: any beneficial subsystem of X changes in the same direction of X, and any harmful subsystem of X changes in the opposite direction of X. The size of the energy changed is equal, but the direction opposite.

Intervention Rule: In the case of virtual disease or economic downturn, the intervening method of intervention is to increase the energy. If the intervening has been done on X, the energy increment (or, increase degree) $|\Delta\varphi(X_S)|$ of the son X_S of X is greater than the energy increment $|\Delta\varphi(S_X)|$ of the mother S_X of X, i.e., the best beneficiary is the son X_S of X. But the energy decrease degree $|\Delta\varphi(X_K)|$ of the prisoner X_K of X is greater than the energy decrease degree $|\Delta\varphi(K_X)|$ of the bane of X, i.e., the worst victim is the prisoner X_K of X.

In the case of real disease or economic overheating, the intervening method of intervention is to decrease the energy. If the intervening has been done on X, the energy decrease degree $|\Delta\varphi(S_X)|$ of the mother S_X of X is greater than the energy decrease degree $|\Delta\varphi(X_S)|$ of the son of X, i.e., the best beneficiary is the mother S_X of X. But the energy increment $|\Delta\varphi(K_X)|$ of the bane K_X of X is greater than the energy increment $|\Delta\varphi(X_K)|$ of the prisoner X_K of X, i.e., the worst victim is the bane K_X of X.

In mathematics, the changing laws are as follows.

1. If $\Delta\varphi(X) = \Delta > 0$, then $\Delta\varphi(X_S) = \rho_1\Delta$, $\Delta\varphi(X_K) = -\rho_1\Delta$, $\Delta\varphi(K_X) = -\rho_2\Delta$, $\Delta\varphi(S_X) = -\rho_2\Delta$;

2. If $\Delta\varphi(X) = -\Delta < 0$, then $\Delta\varphi(X_S) = -\rho_2\Delta$, $\Delta\varphi(X_K) = \rho_2\Delta$, $\Delta\varphi(K_X) = \rho_1\Delta$, $\Delta\varphi(S_X) = -\rho_1\Delta$;

Where $1 \geq \rho_1 \geq \rho_2 \geq 0$. Both ρ_1 and ρ_2 are called intervention reaction coefficients, which are used to represent the capability of intervention reaction. The larger ρ_1 and ρ_2, the better the capability of intervention reaction. The state $\rho_1 = \rho_2 = 1$ is the best state but the state $\rho_1 = \rho_2 = 0$ is the worst state.

This intervention rule is similar to force and reaction in Physics. In other words, if a subsystem of multilateral system V has been intervened, then the energy of subsystem which has neighboring relation changes in the same direction of the force, and the energy of subsystem which has alternate relation changes in the opposite direction of the force. The size of the energy changed is equal, but the direction opposite.

In general, ρ_1 and ρ_2 are decreasing functions of the intervention force Δ since the intervention force $\Delta\varphi(X)$ is easily to transfer all if Δ is small but the intervention force is not easily to transfer all if Δ is large. The energy function of complex system, the stronger the more you use. In order to magnify ρ_1 and ρ_2, should set up a mathematical complex system of the intervention reaction system, and often use it.

Mathematical intervening resistance problem is that such a question, beginning more appropriate mathematiccal intervening method, but is no longer valid after a period. It is because the capability of intervention reaction is bad, i.e., the intervention reaction coefficients ρ_1 and ρ_2 is too small. In the state $\rho_1 = \rho_2 = 1$, any mathematical intervening resistance problem is non-existence but in the state $\rho_1 = \rho_2 = 0$, mathematical intervening resistance problem is always existence. At this point, the paper advocates the essential principle of intervening to avoid mathematical intervening resistance problems.

Self-Protection Rule of a Multilateral System

If there is an intervening force on the subsystem X of a steady multilateral system V which makes the energy $\varphi(X)$ changed by increment $\Delta\varphi(X)$ such that the energies $\varphi(X_S)$, $\varphi(X_K)$, $\varphi(K_X)$ $\varphi(S_X)$ of other subsystems X_S, X_K, K_X, S_X (defined in Theorem 2.4) of V will be changed by the increments $\Delta\varphi(X_S)$, $\Delta\varphi(X_K)$, $\Delta\varphi(K_X)$, $\Delta\varphi(S_X)$, respectively, then can the multilateral system V has capability to protect the worst victim to restore?

It is said that the steady multilateral system has the capability of self-protection if the multilateral system has capability to protect the worst victim to restore. The capability of self-protection of the steady multi-lateral system is said to be better if the multilateral system has capability to protect all victims to restore.

In general, there is an essential principle of self-protection: any harmful subsystem of X should be protected by using the same intervention force but any beneficial subsystem of X should not.

Self-protection Rule: In the case of virtual disease or economic downturn, the intervening method of intervention is to increase the energy. If the intervening has been done on X by the increment $\Delta\varphi(X) = \Delta > 0$, the worst victim is the prisoner X_K of X which has the increment $(-\rho_1\Delta)$. Thus the intervening principle of self-protection is to restore the prisoner X_K of X and the restoring method of self-protection is to increase the energy $\varphi(X_K)$ of the prisoner X_K of X by using the intervention force on X according to the intervention rule. In general, the increase degree is $(\rho_3\Delta)$ where $\rho_3 \leq \rho_1$.

In the case of real disease or economic overheating, the intervening method of intervention is to decrease the energy. If the intervening has been done on X by the increment $\Delta\varphi(X) = -\Delta < 0$, the worst victim is the bane K_X of X which has the increment $(\rho_1\Delta)$. Thus the intervening principle of self-protection is to restore the bane K_X of X and the restoring method of self-protection is to decrease the energy $\varphi(K_X)$ of the bane K_X of X by using the same intervention force on X according to the intervention rule. In general, the decrease degree is $(-\rho_3\Delta)$ where $\rho_3 \leq \rho_1$.

In mathematics, the following self-protection laws hold.

1. If $\Delta\varphi(X) = \Delta > 0$, then the energy of subsystem X_K will decrease the increment $(-\rho_1\Delta)$, which is the worst victim. So the capability of self-protection increases the energy of subsystem X_K by increment

$(\rho_3\Delta)$ where $\rho_3 \leq \rho_1$, in order to restore the worst victim by according to the intervention rule.

2. If $\Delta\varphi(X) = -\Delta < 0$, then the energy of subsystem K_X will increase the increment $(\rho_1\Delta)$, which is the worst victim. So the capability of self-protection decreases the energy of subsystem K_X by increment $(-\rho_3\Delta)$ where $\rho_3 \leq \rho_1$, in order to restore the worst victim by according to the intervention rule.

In general, $0 \leq \rho_3 \leq \rho_1 \leq 1$ the ρ_1 is the intervenetion reaction coefficient. The ρ_3 is called a self-protection coefficient, which is used to represent the capability of self-protection. The larger ρ_3, the better the capability of self-protection. The state $\rho_3 = \rho_1 = 1$ is the best state but the state $\rho_3 = 0$ is the worst state of self-protection. According to the general economy of the protection principle, ρ_3 should be not greater than ρ_1 since the purpose of protection is to restore the victims and not reward the victims.

The self-protection rule can be explained as: the general principle of self-protection subsystem is that the most affected is protected firstly, the protection method and intervention force are in the same way.

In general, ρ_3 is also a decreasing function of the intervention force Δ since the worst victim is easily to restore all if Δ is small but the worst victim is not easily to restore all if Δ is large? The energy function of complex system, the stronger the more you use. In order to magnify ρ_3, should set up an economic society of the self-protection system, and often use it.

Theorem 3.1: Suppose that a steady multilateral system V which has energy function $\varphi(*)$ and capabilities of intervention reaction and self-protection is with intervention reaction coefficients $\rho_1 = \rho_1(\Delta)$ and $\rho_2 = \rho_2(\Delta)$, and with self-protection coefficient $\rho_3 = \rho_3(\Delta)$. If the capability of self-protection wants to restore both subsystems X_K and $K_{X'}$ then the following statements are true.

1. In the case of virtual disease, the treatment method is to increase the energy. If an intervention force on the subsystem X of steady multilateral system V is implemented such that its energy $\varphi(X)$ has been changed by increment $\Delta\varphi(X) = \Delta > 0$, then all five subsystems will be changed finally by the increments as follows:

$$\Delta\varphi(X)_2 = \Delta\varphi(X) + \Delta\varphi(X)_1 = (1 - \rho_2\rho_3)\Delta > 0,$$
$$\Delta\varphi(X_S)_2 = \Delta\varphi(X_S) + \Delta\varphi(X_S)_1 = (\rho_1 + \rho_2\rho_3)\Delta > 0,$$
$$\Delta\varphi(X_K)_2 = \Delta\varphi(X_K) + \Delta\varphi(X_K)_1 = -(\rho_1 - \rho_3)\Delta \le 0,$$
$$\Delta\varphi(K_X)_2 = \Delta\varphi(K_X) + \Delta\varphi(K_X)_1 = -(\rho_2 - \rho_1\rho_3)\Delta \le 0,$$
$$\Delta\varphi(S_X)_2 = \Delta\varphi(S_X) + \Delta\varphi(S_X)_1 = (\rho_2 - \rho_1\rho_3)\Delta \ge 0$$
$$\forall\Delta\varphi(X) = \Delta > 0. \tag{1}$$

2. In the case of real disease, the treatment method is to increase the energy. If an intervention force on the subsystem X of steady multi-lateral system V is implemented such that its energy $\varphi(X)'$ has been changed by increment $\Delta\varphi(X)' = -\Delta < 0$, then all five subsystems will be changed by the increments as follows:

$$\Delta\varphi(X)'_2 = \Delta\varphi(X) + \Delta\varphi(X)'_1 = -(1 - \rho_2\rho_3)\Delta < 0,$$
$$\Delta\varphi(X_S)'_2 = \Delta\varphi(X_S) + \Delta\varphi(X_S)'_1 = -(\rho_2 - \rho_1\rho_3)\Delta \le 0,$$
$$\Delta\varphi(X_K)'_2 = \Delta\varphi(X_K) + \Delta\varphi(X_K)'_1 = (\rho_2 - \rho_1\rho_3)\Delta \ge 0,$$
$$\Delta\varphi(K_X)'_2 = \Delta\varphi(K_X) + \Delta\varphi(K_X)'_1 = (\rho_1 - \rho_3)\Delta \ge 0,$$
$$\Delta\varphi(S_X)'_2 = \Delta\varphi(S_X) + \Delta\varphi(S_X)'_1 = -(\rho_1 + \rho_2\rho_1)\Delta < 0$$
$$\forall\Delta\varphi(X)' = -\Delta < 0. \tag{2}$$

Where the $\Delta\varphi(*)_1$'s and $\Delta\varphi(*)_1'$'s are the increments under the capability of self-protection.

Corollary 3.1: Suppose that a steady multilateral system V which has energy function $\varphi(*)$ and capabilities of intervention reaction and self-protection is with intervention reaction coefficients $\rho_1 = \rho_1(\Delta)$ and $\rho_2 = \rho_2(\Delta)$, and with self-protection coefficient $\rho_3 = \rho_3(\Delta)$. Then the capability of self-protection can make both subsystems X_K and K_X to be restored at the same time, i.e., the capability of self-protection is better, if and only if $\rho_2 = \rho_1\rho_3$ and $\rho_3 = \rho_1$.

Side effects of mathematical intervening problems were the question: in the mathematical intervening process, destroyed the normal balance of non-fall ill subsystem or non-intervention subsystem. By Theorem 3.1 and Corollary 3.1, it can be seen that if the capability of self-protection of the steady multilateral system is better, i.e., the multilateral system has capability to protect all the victims to restore, then a necessary and sufficient condition is $\rho_2 = \rho_1\rho_3$ and $\rho_3 = \rho_1$. General for a complex system of mathematical complex system, the condition $\rho_2 = \rho_1\rho_3$ is easy to meet since it can restore two subsystems by Theorem 3.1, the condition $\rho_3 = \rho_1$ is difficult to meet it only can restore one subsystem by Theorem 3.1. At this point, the paper advocates the principle to avoid any side effects of intervening.

Mathematical Reasoning of Intervening Principle by Using the Neighboring Relations of Steady Multilateral Systems

Intervening principle by using the neighboring relations of steady multilateral systems is "Virtual disease or mathematical virtual disease is to fill his mother but mathematical real disease is to rush down his son". In order to show the rationality of the intervening principle, it is needed to prove the following theorems.

Theorem 3.2: Suppose that a steady multilateral system V which has energy function and capabilities of intervention reaction and self-protection is with intervention reaction coefficients $\rho_1 = \rho_1(\Delta)$ and $\rho_2 = \rho_2(\Delta)$, and with self-protection coefficient $\rho_3 = \rho_3(\Delta)$ satisfying $\rho_2 = \rho_1\rho_3$ and $\rho_3 = \rho_1$. Then the following statements are true.

In the case of virtual disease, if an intervention force on the subsystem X of steady multilateral system V is implemented such that its energy $\varphi(X)$ increases the increment $\Delta\varphi(X) = \Delta > 0$, then the subsystems S_X, X_K and K_X can be restored at the same time, but the subsystems X and X_S will increase their energies by the increments respectively,

$$\Delta\varphi(X)_2 = (1 - \rho_2\rho_3)\Delta\varphi(X)$$
$$= (1 - \rho_2\rho_3)\Delta = (1 - \rho_1^3)\Delta > 0$$

and

$$\Delta\varphi(X_S)_2 = (\rho_1 + \rho_2\rho_3)\Delta\varphi(X)$$
$$= (\rho_1 + \rho_2\rho_3)\Delta = (\rho_1 + \rho_1^3)\Delta > 0,$$

On the other hand, in the case of real disease, if an intervention force on the subsystem X of steady multilateral system V is implemented such that its energy $\varphi(X)'$ decreases, i.e., by the increment $\Delta\varphi(X)' = -\Delta < 0$, the subsystems X_S, X_K and K_X can also be restored at the same time, and the subsystems X and X_S will decrease their energies, i.e., by the increments respectively,

$$\Delta\varphi(X)'_2 = (1 - \rho_2\rho_3)\Delta\varphi(X)'$$
$$= -(1 - \rho_2\rho_3)\Delta = -(1 - \rho_1^3)\Delta < 0$$

and

$$\Delta\varphi(S_X)'_2 = (\rho_1 + \rho_2\rho_3)\Delta\varphi(X)'$$
$$= -(\rho_1 + \rho_2\rho_3)\Delta = -(\rho_1 + \rho_1^3)\Delta < 0,$$

Theorem 3.3: For a steady multilateral system V which has energy function $\varphi(*)$ and capabilities of interveneing reaction and self-protection, assume intervention reaction coefficients are ρ_1 and ρ_2, and let the self-protection coefficient be ρ_3, which satisfy $\rho_2 = \rho_1\rho_3$, $\rho_3 = \rho_1$ and $\rho_1 \geq \rho_0$ where $\rho_0 \approx (<)0.5897545123$ (the following the same) is the solution of $2\rho_1^3 + \rho_1 = 1$. Then the following statements are true.

1. If an intervention force on the subsystem X of steady multilateral system V is implemented such that its energy $\varphi(X)$ has been changed by increment $\Delta\varphi(X) = \Delta > 0$, then the final increment $(\rho_1 + \rho_2\rho_3)\Delta$ of the energy $\varphi(X_S)$ of the subsystem X_S changed is greater than or equal to the final increment $(1 - \rho_2\rho_3)\Delta$ of the energy $\varphi(X)$ of the subsystem X changed based on the capability of self-protection.

2. If an intervention force on the subsystem X of steady multilateral system V is implemented such that its energy $\varphi(X)$ has been changed by increment $\Delta\varphi(X) = -\Delta < 0$, then the final increment $-(\rho_1 + \rho_2\rho_3)\Delta$ of the energy $\varphi(S_X)$ of the subsystem S_X changed is less than or equal to the final increment $-(1 - \rho_2\rho_3)\Delta$ of the energy $\varphi(X)$ of the subsystem X changed based on the capability of self-protection.

Corollary 3.2: For a steady multilateral system V which has energy function $\varphi(*)$ and capabilities of intervening reaction and self-protection, assume intervention reaction coefficients are ρ_1 and ρ_2, and let the self-protection coefficient be ρ_3, which satisfy $\rho_2 = \rho_1\rho_3$, $\rho_3 = \rho_1$ and $\rho_1 < \rho_0$. Then the following statements are true.

1. In the case of virtual disease, if an intervention force on the subsystem X of steady multilateral system V is implemented such that its energy $\varphi(X)$ has been changed by increment $\Delta\varphi(X) = \Delta > 0$, then the final increment $(\rho_1 + \rho_2\rho_3)\Delta$ of the energy $\varphi(X_S)$ of the subsys-

tem X_S changed is less than the final increment $(1-\rho_2\rho_3)\Delta$ of the energy $\varphi(X)$ of the subsystem X changed based on the capability of self-protection.

2. In the case of real disease, if an intervention force on the subsystem X of steady multilateral system V is implemented such that its energy $\varphi(X)$ has been changed by increment $\Delta\varphi(X) = -\Delta < 0$, then the final increment $-(\rho_1 + \rho_2\rho_3)\Delta$ of the energy $\varphi(S_X)$ of the subsystem S_X changed is greater than the final increment $-(1-\rho_2\rho_3)\Delta$ of the energy $\varphi(X)$ of the subsystem X changed based on the capability of self-protection.

By Theorems 3.2 and 3.3 and Corollary 3.2, the intervention method of "Virtual disease or economic downturn is to fill his mother but real disease or economic overheating is to rush down his son" should be often used in case: $\rho_2 = \rho_1\rho_3$, $\rho_3 = \rho_1$ and $\rho_1 \geq \rho_0$ since in this time, $(\rho_1 + \rho_2\rho_3)\Delta \geq (1-\rho_2\rho_3)\Delta$.

Mathematical Reasoning of Intervening Principle by Using the Alternate Relations of Steady Multilateral Systems

Intervening principle by using the alternate relations of steady multilateral systems is "Strong inhibition of the same time, support the weak". In order to show the rationality of the intervening Principle, it is needed to prove the following theorems.

Theorem 3.4: Suppose that a steady multilateral system V which has energy function $\varphi(*)$ and capabilities of intervention reaction and self-protection is with intervention reaction coefficients $\rho_1 = \rho_1(\Delta)$ and $\rho_2 = \rho_2(\Delta)$, and with self-protection coefficient $\rho_3 = \rho_3(\Delta)$. Then the following statements are true.

Assume there are two subsystems X and X_K of V with an alternate relation such that X encounters virtual disease, and at the same time, X_K befalls real disease. If an intervention force on the subsystem X of steady multilateral system V is implemented such that its energy $\varphi(X)$ has been changed by increment $\Delta\varphi(X) = \Delta > 0$, and at the same time, another intervention force on the subsystem X_K of steady multilateral system V is also implemented such that its energy $\varphi(X_K)$ has been changed by increment $\Delta\varphi(X_K)' = -\Delta < 0$, then all other subsystems: $S_{X'}$, K_X and X_S can be restored at the same time, and the subsystems X and X_K will increase and decrease their energies by the same size but the direction opposite, i.e., by the increments respectively,

$$\Delta\varphi(X)_3 = (1 - \rho_2\rho_3)\Delta\varphi(X)$$
$$= (1 - \rho_2\rho_3)\Delta = (1 - \rho_1^3)\Delta > 0$$

and

$$\Delta\varphi(X_K)_3 = (1 - \rho_2\rho_3)\Delta\varphi(X_K)$$
$$= -(1 - \rho_2\rho_3)\Delta = -(1 - \rho_1^3)\Delta < 0.$$

Assume there are two subsystems X and K_X of V with an alternate relation such that X encounters real disease, and at the same time, K_X befalls virtual disease. If an intervention force on the subsystem X of steady multilateral system V is implemented such that its energy $\varphi(X)$ has been changed by increment $\Delta\varphi(X) = -\Delta < 0$, and at the same time, another intervention force on the subsystem K_X of steady multilateral system V is also implemented such that its energy $\varphi(K_X)$ has been changed by increment $\varphi(K_X) = \Delta > 0$, then all other subsystems: $S_{X'}$, X_K and X_S can be restored at the same time, and the subsystems X and K_X will decrease and increase their energies by the same size but the direction opposite, i.e., by the increments respectively,

$$\Delta\varphi(X)_3 = (1 - \rho_2\rho_3)\Delta\varphi(X)$$
$$= -(1 - \rho_2\rho_3)\Delta = -(1 - \rho_1^3)\Delta < 0$$

and

$$\Delta\varphi(K_X)_3 = (1 - \rho_2\rho_3)\Delta\varphi(K_X)$$
$$= (1 - \rho_2\rho_3)\Delta = (1 - \rho_1^3)\Delta > 0.$$

By Theorems 3.3 and 3.4 and Corollary 3.2, the method of "Strong inhibition of the same time, support the weak" should be used in case: $\rho_2 = \rho_1\rho_3$, $\rho_3 = \rho_1$ and $\rho_1 < \rho_0$ since $(\rho_1 + \rho_2\rho_3)\Delta < (1 - \rho_2\rho_3)\Delta$.

RATIONALITY OF INTERVENING PRINCIPLE OF TRADITIONAL CHINESE MATHEMATICS AND "YIN YANG WU XING" THEORY

Chinese Traditional Mathematics and "Yin Yang Wu Xing" Theory

Ancient Chinese "Yin Yang Wu Xing" [38] Theory has been surviving for several thousands of years without dying out, proving it reasonable to some extent. If we regard ~ as the same category, the neighboring relation → as beneficial, harmony, obedient, loving, etc. and the alternate relation ⟹ as harmful, conflict, ruinous, killing, etc., then the above defined stable logic analysis model is similar to the logic architecture of reasoning of "Yin Yang Wu Xing". Both "Yin" and "Yang" mean that there are two opposite relations in the world: harmony or loving → and conflict or killing ⟹, as well as a general equivalent category ~. There is only one of three relations ~, → and ⟹ between every two objects. Everything X makes something ($X_S \neq \varnothing$), and is made by something ($S_X \neq \varnothing$); everything restrains something ($X_K \neq \varnothing$), and is restrained by something ($K_X \neq \varnothing$); i.e., one thing overcomes another thing and one thing is overcome by another thing. The ever changing world V, following the relations: ~, → and ⟹, must be divided into five categories by the equivalent relation ~, being called "Wu Xing": wood (X), fire (X_S), soil (X_K), metal (K_X), and water (S_X). The "Wu Xing" is to be "neighbor is friend": wood (X) → fire (X_S) → soil (X_K) → metal (K_X) → water (S_X) → wood (X), and "alternate is foe": wood (X) ⇒ soil (X_K) ⇒ water (S_X) ⇒ fire (X_S) ⇒ metal (K_X) ⇒ wood (X). In other words, the ever changing world must be divided into five categories:

$$V = X + X_S + X_K + K_X + S_X$$

Satisfying

$$X \to X_S \to X_K \to K_X \to S_X \to X$$

And

$$X \Rightarrow X_K \Rightarrow S_X \Rightarrow X_S \Rightarrow K_X \Rightarrow X$$

Where elements in the same category are equivalent to one another. We can see, from this, the ancient Chinese "Yin Yang Wu Xing" theory is a reasonable logic analysis model to identify the stability and relationship of complex mathematical systems.

Image mathematics firstly uses the verifying relationship method of "Yin Yang Wu Xing" Theory to explain the relationship between mathematical complex system and environment. Secondly, based on "Yin Yang Wu Xing" Theory, the relations of development processes of mathematical complex system can be shown by the neighboring relation and alternate relation of five subsets. Then a normal mathematical complex system can be shown as a steady multilateral system in which there are the loving relation and the killing relation and the liking relation. The loving relation in image mathematics can be explained as the neighboring relation, called "Sheng (生)". The killing relation in image mathematics can be explained as the alternate relation, called "Ke (克)". The liking relation can be explained as the equivalent relation, called "Tong-Lei (同类)". Constraints and conversion between five subsets are equivalent to the two kinds of triangle reasoning. So a normal mathematical complex system can be classified into five equivalence classes corresponding five mathematical indexes, respectively.

For example, in image mathematics, a mathematical complex system is similar to a human body. A mathematical index system of normal complex system following the "Yin Yang Wu Xing" Theory was classified into five equivalence classes as follows [29]:

Consider a complex system, its input $x_1,...,x_m,\omega$ and output y can be written as

$$y = f\left(x_1,\cdots,x_m,\omega\right) = g\left(x_1,\cdots,x_m\right) + \varepsilon_g$$

$$g\left(x_1,\cdots,x_m\right) = E\left(f\left(x_1,\cdots,x_m,\omega\right)\middle|x_1,\cdots,x_m\right),$$

$$\varepsilon_g = f\left(x_1,\cdots,x_m,\omega\right) - E\left(f\left(x_1,\cdots,x_m,\omega\right)\middle|x_1,\cdots,x_m\right),$$

(2)

Where not only all output functions

$y = f = f(x_1,...,x_m,\omega)$, $g(x_1,...,x_m)$ and ε_g are not known, but also the input variables $x_1,...,x_m,\omega$ are not known. The problem is called model-free.

The inputs $x_1,...,x_m$ are called controllable if they are observed and controlled by human. So $g(x_1,...,x_m) = E(f(x_1,...,x_m,\omega)|x_1,...,x_m)$ can be observed if the controllable inputs $x_1,...,x_m$ can be choose.

The input ω is called uncontrolled if it is not observed or controlled by human. So the freedom model error

$$\varepsilon_g = f\left(x_1,\cdots,x_m,\omega\right) - E\left(f\left(x_1,\cdots,x_m,\omega\right)\middle|x_1,\cdots,x_m\right)$$

Cannot be assumed. But we can show the following properties:

$$E\left(\varepsilon_g \middle| x_1, \cdots, x_m\right) = E\left(f\left(x_1, \cdots, x_m, \omega\right) \middle| x_1, \cdots, x_m\right)$$
$$- E\left[E\left(f\left(x_1, \cdots, x_m, \omega\right) \middle| x_1, \cdots, x_m\right) \middle| x_1, \cdots, x_m\right] = 0,$$

$$Var\left(\varepsilon_g^2 \middle| x_1, \cdots, x_m\right) =: \sigma_{x_1 \cdots x_m}^2 \geq 0,$$

$$Cov\left(\left(g\left(x_1, \cdots, x_m\right), \varepsilon_{g(x_1 \cdots x_m)}\right) \middle| x_1, \cdots, x_m\right)$$
$$= E\left(\left(g - Eg\right)\left(\varepsilon_g - E\varepsilon_g\right) \middle| x_1, \cdots, x_m\right) = 0.$$

The condition is not hypothesis since they can be obtained if f makes the calculation meaningful, such as, if f is continuous.

In general, we can consider the inputs $x_1, \dots, x_m, \omega$ as independent random variables with continuous distributions $F_1(x_1), \dots, F_m(x_m), F_\omega(\omega)$, respectively, since the inputs $x_1, \dots, x_m$ are controllable which can be selected independently by human under some similar conditions. If not independent, by factor analysis can select orthogonal factor. This operation of experiment under some similar conditions is equivalent to the uncontrolled input ω is independent of the controllable input $x_1, \dots, x_m$ since $Cov(g(x_1, \dots, x_m), \varepsilon_g) = 0$ This operation of experiment that the inputs $x_1, \dots, x_m$ can be selected independently by human is equivalent to that the inputs $x_1, \dots, x_m$ are independent random variables one another. For example, take $x_1, \dots, x_m$ based on an orthogonal array since the orthogonality is equivalent to independence for discrete random variables and continuous random variables can be in a discrete random variable approximation [8].

In this case, it is well known that the inputs are independent random variables with the same continuous distribution $U(0,1)$,

$$u_1 = F_1(x_1), \cdots, u_m = F_m(x_m), u_\omega = F_\omega(\omega)$$

Assume

$$v_1 = a_1 + u_1(b_1 - a_1), \cdots, v_m = a_m + u_m(b_m - a_m),$$
$$v_\omega = a_\omega + u_\omega(b_\omega - a_\omega),$$

Where $b_j > a_j, j = 1, \ldots, m, b_\omega > a_\omega$. Then the inputs $v_1, \ldots, v_m, v_\omega$ are independent random variables with continuous distributions $U(a_1, b_1), \ldots, U(a_m, b_m), U(a_\omega, b_\omega)$.

In this case, the function

$$h(v_1, \cdots, v_m, v_\omega) = f\left(F_1^{-1}\left(\frac{v_1 - a_1}{b_1 - a_1}\right), \cdots, \right.$$

$$\left. F_m^{-1}\left(\frac{v_m - a_m}{b_m - a_m}\right), F_\omega^{-1}\left(\frac{v_\omega - a_\omega}{b_\omega - a_\omega}\right)\right)$$

can replace f as a new system function since each of both h and f are all not known. Therefore, without loss of generality, we always consider $x_1, \ldots, x_m, \omega$ as independent random variables with continuous distributions

$$U(a_1, b_1), \cdots, U(a_m, b_m), U(a_\omega, b_\omega).$$

On the other hand, the function f is considered as continuous, in order to ensure that condition expectations and partial derivative of existence and make the conventional mathematics method has significance.

To the complex system f, we need to decide an energy goal t, make y more close to target, the greater the function of the system. In general, the Target t is the maximum energy of y.

Image mathematics in TCMath considers the complex system stability problem, because the core problem of any complex system is stability. The stability can only through the fixed program to observe to do a test or experiment since the function f is not known. In general, the human wants to find a testing or experimental center $x^0 = (x_1^0, ..., x_m^0)$ and testing or experimental tolerance $\Delta x = (\Delta x_1, ..., \Delta x_m)$, $\Delta_{max} = max(\Delta x_1, ..., \Delta x_m)$, for the observed function

$$g(x_1, \cdots, x_m) = E\left(f(x_1, \cdots, x_m, \omega) \middle| x_1, \cdots, x_m \right)$$

Under some similar conditions, such that

$$\left(Eg(x_1, \cdots, x_m) - t \right)^2$$

$$= \left(Eg(x_1^0, \cdots, x_m^0) - t \right)^2 + O(\Delta_{max}) \rightarrow min;$$

1. $E(y - t)^2 \rightarrow min;$

Where the controllable inputs $x_1, ..., x_m$ are independent random variables and

$$x_j \sim U(a_j, b_j) = U(x_j^0 - \Delta x_j, x_j^0 + \Delta x_j), \, j = 1, \cdots, m.$$

In order to solve the stability problem, we get easily the following theorem [29]:

Theorem 4.1: Suppose that f is continuous and

$$g = g(x_1, \cdots, x_m) = E\left(f(x_1, \cdots, x_m, \omega) \middle| x_1, \cdots, x_m \right).$$

Then

$$E(y-t)^2 = (Eg-t)^2 + Var(g) + Var(\varepsilon_g),$$

Where $Var(\varepsilon_g) = E\sigma_{x_1,\dots,x_m}^2$

1) In image mathematics, the index

$U_g^2 = Var(g(x_1, \dots, x_m))$ is called Distortion Degree. It belongs to the wood (X) subsystem of the complex system f since it cognizes the structure function g of the complex system f which is the beginning or birth stage of all things, just like in the Spring of a year. In mathematics,

$$U_g^2 = Var\left(g(x_1, \cdots, x_m) \right)$$

$$= \sum_{j=1}^{m} \left(\frac{\partial g(x^0)}{\partial x_j} \right)^2 Var(x_j) + o(\Delta_{max}) = U_g^2(\Delta_{max})$$

Where $\Delta_{max} = max(\Delta x_1, \dots, \Delta x_m)$ It is thought as the wood (X) image of generally complex system f since it is an objective constant independent of human observations, expressing birth, although the maximum and controllable experimental tolerance Δ_{max} can be choose by human.

The index $\varphi(X) = Var(g(x_1, \dots, x_m))^{-1} = U_g^2$ can be taken as the energy function of the subsystem wood (X) since the smaller the distortion degree U_g^2, the better the stability of the complex system f.

Although the distortion degree U_g^2 is unknown for freedom model speaking, it can be easily estimated by using orthogonal arrays $L_n(p_1,...,p_m) = (a_{ij})_{n \times m}$,

$0 \le a_{ij} \le p_j - 1, j = 1,...,m$ (See [29]). For example, take

$$x_{ij} = x_j\left(a_{ij}\right) = x_j^0 + C_j\left(a_{ij}\right)\Delta x_j$$

Where

$C_j(a_{ij}) = -1 + 2/(p_j - 1)$ then $Var(x_j) = W(p_j)\Delta x_j^2$, where

$$W\left(p_j\right) = \frac{1}{p_j}\sum_{k=0}^{p_j-1} C_j(k)^2 .$$

And for the experiment data $y_i = f(x_{i1}, x_{im}, \omega_{i\omega})$, $i = 1,...,n$, we have

$$\frac{\widehat{\partial g\left(x^0\right)}}{\partial x_j} = \frac{\dfrac{1}{p_j}\sum_{k=0}^{p_j-1} C_j(k)\hat{\mu}_{jk}}{W\left(p_j\right)\Delta x_j},$$

$$\hat{\mu}_{jk} \triangleq \frac{1}{r_{jk}}\sum_{s\in H_{jk}} y_s, \quad r_{jk} \triangleq \left|H_{jk}\right|,$$

Where $H_{jk} = \{s : a_{sj} = k, s = 1,...,n\}$. Thus, we can obtained the estimation of the distortion degree $U_g^2 = Var(g(x_1,...,x_m))$ as

$$\widehat{U_g^2} = \widehat{Var(g)} = \sum_{j=1}^{m} \frac{\left(\frac{1}{P_j} \sum_{k=0}^{P_j-1} C_j(k) \widehat{\mu}_{jk} \right)^2}{W(P_j)}.$$

The $\Delta_{max} = max(\Delta x_1, ..., \Delta x_m)$ smaller, more exact estimate.

2) The index $\gamma_f^2 = Var(f) = Var(g) + Var(\varepsilon_g)$ is called Disturb Degree. It belongs to the fire (X_S) subsystem of the complex system f since it controls the fluctuations of the complex system f which is the development and growth stage of all things, just like in the summer of a year. In mathematics,

$$\gamma_f^2 = Var\left(f\left(x_1, \cdots, x_m\right)\right)$$

$$= \sum_{j=1}^{m} \left(\frac{\partial g\left(x^0\right)}{\partial x_j} \right)^2 Var\left(x_j\right) + E\sigma_{x_1,...,x_m}^2 + o\left(\Delta_{max}\right)$$

$$= \gamma_g^2\left(\Delta_{max}\right)$$

Where $\Delta_{max} = max(\Delta x_1, ..., \Delta x_m)$ It is thought as the fire (X_S) image of generally complex system f since it is also an objective constant independent of human observations, expressing growth, although the maximum and controllable experimental tolerance Δ_{max} can be choose by human. The index $\varphi(X_S) = \gamma_f^{-2}$ can be taken as the energy function of the subsystem fire (X_S) since the smaller the disturb degree γ_f^2, the better the stability of the complex system f.

Although the disturb degree γ_f^2 is unknown for freedom model speaking, it can be easily estimated by using experiment data $y_1, ..., y_n$ of orthogonal arrays (see [29]). For example, a good estimation of $\gamma_f^2 = \mathrm{Var}(f)$ is the data standard variance

$$\widehat{\gamma_f^2} = \widehat{\mathrm{Var}(f)} = S_n^2 = \frac{1}{n-1}\sum_{s=1}^{n}(y_s - \bar{y})^2 ,$$

Where

$$\bar{y} = \frac{1}{n}\sum_{s=1}^{n}y_s .$$

3) The index $\eta_f = (Ef)^2 / \mathrm{Var}(f)$ is called Information Decomposition Ratio (or Signal to Noise Ratio). It belongs to the soil (X_K) subsystem of the complex system f since it makes the coordination of the center and fluctuation in the complex system f which is the continuous development and combined stage of all things, just like in the Long-Summer of a year. In mathematics,

$$\eta_f = (Ef)^2 \big/ \mathrm{Var}\big(f(x_1,\cdots,x_m)\big)$$

$$= \frac{(Eg)^2}{\displaystyle\sum_{j=1}^{m}\left(\frac{\partial g(x^0)}{\partial x_j}\right)^2 \mathrm{Var}(x_j) + E\sigma_{x_1,\ldots,x_m}^2 + o(\Delta_{max})}$$

$$= \eta_g(\Delta_{max})$$

Where $\Delta_{max} = \max(\Delta x_1,\ldots,\Delta x_m)$ It is thought as the soil $(X_{K'})$ image of generally complex system f since it is an objective constant independent of human observations, expressing combined, although the maximum and controllable experimental tolerance Δ_{max} can be choose by human. The index $\varphi(X_K) = \eta_f$ can be taken as the energy function of the subsystem soil (X_K) since the bigger the information decomposition ratio η_f, the better the stability of the complex system f.

Although the information decomposition ratio η_f is unknown for freedom model speaking, it can be easily estimated by using the experimental data $y_1,...,y_n$ of orthogonal arrays (see [29]). For example, a good estimation of η_f is

$$\widehat{\eta}_f = \frac{n\overline{y}^2}{S_n^2}, \quad S_n^2 = \frac{1}{n-1}\sum_{s=1}^{n}\left(y_s - \overline{y}\right)^2, \quad \overline{y} = \frac{1}{n}\sum_{s=1}^{n}y_s.$$

Note: In data analysis situation, often taking $\eta_f = nE\overline{y}^2 / \gamma_g^2$ instead of $\eta_f = (Ef)^2 / Var(f)$. Limited to data observation point of view, they are in the statistical meaning equivalent, but the former is more advantageous to the statistical analysis.

4) The index $\rho_f^2 = (Ef - t)^2$ is called Deviation Degree. It belongs to the metal (K_X) subsystem of the complex system f since it makes function characteristics (or data center and the expected goal deviation) in the complex system f which is the getting-results and accepted stage of all things, just like in the autumn of a year. In mathematics,

$$\rho_f^2 = \left(Ef - t\right)^2 = \left(Eg - t\right)^2$$

$$= \left(g\left(x^0\right) - t\right)^2 + o\left(\Delta_{max}\right) = \rho_g^2\left(\Delta_{max}\right)$$

Where

$$Ey = Ef = Eg = g\left(x^0\right) + o\left(\Delta_{max}\right).$$

It is thought as the metal (K_X) image of a generally complex system f since it is an objective constant independent of human observations expressing accepted, although the controllable inputs $x_1,...,x_m$ of the observed function

$$g(x_1,\cdots,x_m) = E\left(f(x_1,\cdots,x_m,\omega) \middle| x_1,\cdots,x_m\right)$$

can be choose by human. The index $\varphi(K_X) = \rho_f^{-2}$ can be taken as the energy function of the subsystem metal (K_X) since the smaller the deviation degree ρ_f^2, the better the stability of the complex system f.

Although the deviation degree ρ_f^2 is unknown for freedom model speaking, it can be easily estimated by using the experimental data $y_1,...,y_n$ of orthogonal arrays (see [29]). For example, a good estimation of ρ_f^2 is

$$\widehat{\rho_f^2} = \left(\overline{y} - t\right)^2, \quad \overline{y} = \frac{1}{n}\sum_{s=1}^{n} y_s.$$

5) The index $R_f^2 = E(y-t)^2$ is called Risk Function (or Risk, or Loss Function). It belongs to the water (S_X) subsystem of the complex system f since it makes the expected value of loss $L(y,t) = (y-t)^2$ between each data y and the goal t in the complex system f which is the risk and hiding stage of all things, just like in the winter of a year. The best condition is $R_f^2 = E(y-t)^2 = 0$, but generally this is very hard to achieve, because right now all the data are equal to the target t and the system f is a simple system $f = t$. In mathematics,

$$R_f^2 = E\left(y-t\right)^2$$
$$= \rho_g^2\left(\Delta_{max}\right) + U_g^2\left(\Delta_{max}\right) + E\sigma_{x_1,\cdots,x_m}^2$$
$$= R_g^2\left(\Delta_{max}\right).$$

Where $E\sigma^2_{x_1,\cdots,x_m} = E(\varepsilon^2_g \mid x_1,\cdots,x_m)$ It is thought as the water (S_x) image of generally complex system f since it is an objective constant independent of human observations, expressing hiding, although the observed function

$$g\left(x_1,\cdots,x_m\right) = E\left(f\left(x_1,\cdots,x_m,\omega\right)\middle| x_1,\cdots,x_m\right)$$

can be choose by human. The index $\varphi(S_x) = R^{-2}_f$ can be taken as the energy function of the subsystem water (S_x) since the smaller the risk R^2_f, the better the stability of the complex system f.

Although the risk R^2_f is unknown for freedom model speaking, it can be easily estimated by using the experimental data $y_1,...,y_n$ of orthogonal arrays (see [29]). For example, a good estimation of R^2_f is

$$\widehat{R^2_f} = \frac{1}{n}\sum_{s=1}^{n}\left(y_s - t\right)^2.$$

In image mathematics, each of the rows of orthogonal arrays is called one gua (卦) which is independent of the unknown system function f (model-free). The state space V of a multilateral system $(V,\mathfrak{R})$ based the unknown system function f is the set of mathematical indexes of the unknown system function f. By Theorem 2.4, the set V can be divided into five categories. Corresponding to every kind of five categories, each of mathematical indexes of the unknown system function f is called an image. Each of images must shows the complex system certain characteristics, such as, wood, fire, soil, metal, water.

For given each of the gua (卦) or each the rows of orthogonal arrays and for any unknown continuous system function f, it can be proved easily that each of true image mathematical indexes above can be obtained if the experiments or observations can be repeated (law of averages of great numbers). The way or calculation method of the image indexes is also independent of the unknown system function f (model-free).

In image mathematics, each of mathematical indexes represents an "Axiom system", called a class. For each of classes, all theories and methods are in order to increase the energy of class or to make the corresponding mathematical index becomes to better. There are the loving and hating (or killing) relations among all images or mathematical indexes of classes. Generally speaking, close is love, alternate is hate.

In every category of internal, think that they are equivalent relationship, between each two of their elements there is a force of similar material accumulation of each other. It is because their pursuit of the goal is the same, i.e., follows the same "Axiom system". It can increase the energy of the class if they accumulate together. All of nature material activity follows the principle of maximizing so energy. In general, the force of similar material accumulation of each other is smaller than the loving force or the killing force in a stable complex system. The stability of any complex system first needs to maintain the equilibrium of the killing force and the loving force. For a stable complex system, if the killing force is large, i.e., p_1 becomes larger, then the loving force is large and the force of similar material accumulation of each other is also large. They can make the complex system more stable. If the killing force is small, i.e., p_1 becomes smaller, then the loving force is small and the force of similar material accumulation of each other is also small. They can make the complex system becoming unstable.

It has been shown in Theorems 2.1 - 2.4 that the classification of five subsets is quite possible based on the mathematical logic. As for the characteristics of the five subsets is rational or not, it is need more research work. It has been also shown in Theorems 3.1 - 3.4 that the logical basis of image mathematics is a steady multilateral system.

The vigor energy (or, Chi, Qi) of image mathematics means the energy function in a steady multilateral system.

There are two kinds of mathematical diseases in image mathematics: Mathematical real disease and mathematical virtual disease. They generally means the subsystem is abnormal, its energy is too high for mathematical real disease or too low for mathematical virtual disease.

The intervening method of image mathematics is to "xie Chi" which means to rush down the energy if a mathematical real disease is treated, or to "bu Chi" which means to fill the energy if a mathematical virtual disease is treated. Like intervening the subsystem, decrease when the energy is too high, increase when the energy is too low.

Both the capability of intervention reaction and the capability of self-protection of the multilateral system are equivalent to the Immunization of image mathematics. This capability is really existence for a mathematical complex system. Its target is to protect other mathematical subsystem while treating one mathematical subsystem. It is because if the capability is not existence, then $\rho_1 = \rho_2 = \rho_3 = 0$. In this time, the energy of the system will be the sum of energy of each part. Thus the mathematical system will be a simple mathematical system which is not what we consider range.

Intervening Principle if Only One Subsystem of the Mathematical Complex System Falls Ill

If we always intervene the abnormal subsystem of the mathematical complex system directly, the intervention method always destroy the balance of the mathematical complex system because it is having strong side effects to the mother or the son of the subsystem which may be non-disease of mathematical subsystem or non-intervened subsystem by using Theorem 3.2. The intervening method also decrease the capability of intervention reaction because the method which don't use the capability of intervention reaction makes the ρ_1 and ρ_2 near to 0. The state $\rho_1 = \rho_2 = 0$ is the worst state of the mathematical complex system, namely mathematical intervenetion failure. On the way, the mathematical intervening resistance problem will be occurred since any mathematical intervening method is possible too little for some small ρ_1 and ρ_2.

But, by Corollary 3.2, it will even be better if we intervene sub-system X itself directly $p_2 = p_1 p_3$, $p_3 = p_1$ and $p_1 < p_0$. In this case: $p_1 + p_2 < 0.9375648971$. It can be explained that if a multilateral system which has a poor capability of intervention reaction, then it is better to intervene the subsystem itself directly than indirectly. But similar to above, the intervention method is always to destroy the balance of multilateral systems such that there is at least one of side effects occurred. And the intervention method also have harmful to the capability of intervention reaction making the mathematical intervene-ing resistance problem also occurred by Theorem 3.2. Therefore the intervention method directly can be used in case $p_1 < p_0$ but should be used as little as possible.

If we always intervene the abnormal subsystem of the mathematic complex system indirectly, the intervention method can be to maintain the balance of the mathematical complex system because it has not any side effects to all other subsystems which are not both the mathemati-cal disease subsystem and the mathematical intervened subsystem by using Theorem 3.2. The intervening method also increase the capabil-ity of intervention reaction because the method of using the interven-tion reaction makes the p_1 and p_2 near to 1. The state $p_1 = p_2 = 1$ is the best state of the mathematical complex system. On the way, it is almost none mathematical intervening resistance problem since any math-ematical intervening method is possible good for some large p_1 and p_2.

For example, in China, many mathematical engineers generally only care about risk R_f^2 (or, income, water (S_X)) and interfere γ_f^2 (or, disturb, fire (X_S)) as two mathematical indexes which have the killing relations for a complicated engineering problem based on the freedom model (3). They frequently used method is: If only the risk R_f^2 (i.e., water (S_X)) is not normal (or, the bigger, virtual disease), then they reduce deviation degree ρ_f^2 (i.e., increase the energy $\varphi(K_X) = \rho_f^{-2}$ of metal (K_X)) with linear optimization method since metal (K_X) is the mother of water (S_X); If only the interfere degree γ_f^2 (i.e., fire (X_S)) is not normal (or, the bigger, vir-tual disease), then they reduce distortion degree U_f^2 (i.e., increase the energy $\varphi(X) = U_f^{-2}$ of wood (X)) of system function f with the method of adjusting stable center x^0 of the system function since wood (X) is the mother of fire (X_s) (see [29]). The idea is precisely "Virtual disease is to fill his mother" if one subsystem of mathematical complex system falls virtual ill.

All in all, the mathematical complex system satisfies the intervention rule and the self-protection rule. It is said a healthy mathematical complex system when the intervention reaction coefficient ρ_1 satisfies $\rho_1 \geq \rho_0$. In logic and practice, it's reasonable $\rho_1 + \rho_2$ near to 1 if the input and output in a complex system is balanced, since a mathematical output subsystem is absolutely necessary social other subsystems of all consumption. In case: $\rho_1 + \rho_2 = 1$, all the energy for intervening mathematical complex subsystem can transmit to other mathematical complex subsystems which have neighboring relations or alternate relations with the intervening mathematical complex subsystem. The condition $\rho_1 \geq \rho_0$ can be satisfied when $\rho_2 = \rho_1\rho_3$ and $\rho_3 = \rho_1$ for a mathematical complex system since $\rho_1 + \rho_2 = 1$ implies $\rho_1 = (\sqrt{5}-1)/2 \approx 0.618 > \rho_0$ And $\rho_2 = 1-(\sqrt{5}-1)/2 \approx 0.382$. If this assumptions is set up then the intervening principle: "Real disease is to rush down his son and virtual disease is to fill his mother" based on "Yin Yang Wu Xing" Theory in image mathematics, is quite reasonable. But, in general, the ability of self-protection often is insufficient for an usual mathematical complex system, i.e., ρ_3 is small. A common standard is $\rho_3 = \dfrac{1-\rho_1}{2\rho_2} \approx \dfrac{1}{2}$, i.e., there is a principle which all losses are bear in mathematical complex system. Thus the general condition often is $\rho_1 \approx 0.618 \geq \rho_3 \approx 0.5 \geq \rho_2 \approx 0.382$. Interestingly, they near to the golden numbers.

On the other hand, in image mathematics, mathematiccal real disease and mathematical virtual disease have their reasons. Mathematical real disease is caused by the born subsystem and mathematical virtual disease is caused by the bear subsystem. Although the reason cannot be proved easily in mathematics or experiments, the intervening method under the assumption is quite equal to the intervening method in the intervention indirectly. It has also proved that the mathematical intervening princeple is true from the other side.

Intervening Principle If Only Two Subsystems with the Loving Relation of the Mathematical Complex System Encounter Sick

Suppose that the two subsystems X and X_S of the mathematic complex system are abnormal (mathematical virtual disease or mathematical real disease). In the mathematical complex system of relations between two noncompatible with the constraints, by Theorem 3.2, only two situations may occur:

1. X encounters mathematical virtual disease, and at the same time, X_S befalls mathematical virtual disease, i.e., the energy of X is too low and the energy of X_S is also too low. It is because X bears X_S. The mathematical virtual disease causal is X.

2. X encounters mathematical real disease, and at the same time, X_S befalls mathematical real disease, i.e., the energy of X is too high and the energy of X_S is also too high. It is because X_S is born by X. The mathematical virtual disease causal is X_S.

It can be shown by Theorem 3.3 that when intervenetion reaction and self-protection coefficients satisfy $\rho_2 = \rho_1\rho_3$, $\rho_3 = \rho_1$ and $\rho_1 \geq \rho_0$, if one wants to treat the abnormal subsystems X and X_S, then,

1. For mathematical virtual disease, the one should intervene subsystem X directly by increasing its energy. It means "mathematical virtual disease is to fill his mother" because the mathematical virtual disease causal is X;

2. For mathematical real disease, the one should intervene subsystem X_S directly by decreasing its energy. It means "mathematical real disease is to rush down his son" because the mathematical real disease causal is X_S.

For example, in China, many factories and enterprises generally only care about cost R_f^2 (or, risk, water (S_x)) and quality ρ_f^2 (or, deviation degree, metal (K_x)) as two mathematical indexes which have the loving relation for a complicated economical problem based on the freedom model (3). They frequently used method is: If both cost and quality are all not normal (or, virtual diseases, i.e., high cost and poor quality), then they reduce deviation degree ρ_f^2 (i.e., improve quality, increase the energy $\varphi(K_X) = \rho_f^{-2}$ of metal (K_X)) with linear optimization method since metal (K_X) is the mother of water (S_X) (see [29]). The idea is precisely "Virtual disease is to fill his mother" if one subsystem of mathematical complex system falls virtual ill.

The intervention method can be to maintain the balance of the mathematical complex system because only two mathematical virtual disease subsystems are treated, by using Theorem 3.2, such that there is not any side effect for all other subsystems. And the intervention method can also be to enhance the capability of intervention reaction because the method of using intervention reaction makes the ρ_1 and ρ_2 greater and near to 1. The state $\rho_1 = \rho_2 = 1$ is the best state of the mathematical complex system. On the way, it almost have none mathematical intervening resistance problem since any mathematical intervening method is possible good for some large ρ_1 and ρ_2.

Intervening Principle If Only Two Subsystems with the Killing Relation of the Mathematical Complex System Encounter Sick

Suppose that the subsystems X and X_K of a mathematical complex system are abnormal (real disease or virtual disease). In the mathematical complex system of relations between two non-compatible with the constraints, only a situation may occur: X encounters mathematical virtual disease, and at the same time, X_K befalls mathematical real disease, i.e., the energy of X is too low and the energy of X_K is too high. The disease is serious because the X_K has harmed the X by using the method of incest, i.e. damaged the king relation of X and X_K.

It can be shown by Theorems 3.3 and 3.4 that when intervention reaction and self-protection coefficients satisfy $\rho_2 = \rho_1\rho_3$, $\rho_3 = \rho_1$ and $\rho_1 < \rho_0$, if one wants to treat the abnormal subsystems X and X_K, the one should intervene subsystem X directly by increasing its energy, and at the same, intervene subsystem X_K directly by decreasing its energy. It means that "Strong inhibition of the same time, support the weak".

For example, in China, many sociologists generally only care about social and economic benefits $\varphi(S_X) = R_f^{-2}$ (or social benefits, water (S_X)) and social equity extent $\varphi(X_K) = \eta_f$ (or, information decomposition ratio, soil (X_K)) with two mathematical killing indexes for a complicated social system based on the freedom model (3). If both social benefit $\varphi(S_X) = R_f^{-2}$ and social equity extent $\varphi(X_K) = \eta_f$ are not normal, often appear serious problem is that social benefit $\varphi(S_X) = R_f^{-2}$ is too good (or, R_f^2 too low, real disease) but social equity extent $\varphi(X_K) = \eta_f$ is too bad (or, η_f too low, i.e., virtual disease). The disease is serious because the X_K has been harmed by the S_X with the method of incest such that the soil (X_K) cannot kill water (S_X). For the serious disease, many mathematical sociologists of China now frequently used method is: to increase social equity extent η_f (or, to increase the information decomposition ratio η_f which is the energy of the soil (X_K)) but to decrease social and economic benefits $\varphi(S_X) = R_f^{-2}$ (i.e., to decrease the energy of water (S_X)) at the same time since soil (X_K) is the bane of water (S_X). The idea is "Strong inhibition of the same time, support the weak" if X_K falls virtual disease and at the same time, S_X befalls real disease.

For another example, thirty years ago, China's social coordination function $\varphi(X_K) = \eta_f$ is very good, the complex system is rife with average socialist, but both social benefit $\varphi(S_X) = R_f^{-2}$ and social structure $\varphi(X) = U_f^{-2}$ are all very poor. In addition, the mathematical complex system is not rich. In other words, both S_X and X fall virtual diseases and at the same time, X_K befalls real disease based on the freedom model (3). The disease is very serious because not only the wood (X) has been harmed by the soil (X_K) with the method of incest such that wood (X) cannot kill soil (X_K), but also there are 3 subsystems falling ill. Generally speaking, there are three or more than three of disease of the subsystem of a complex system, it is very difficult to cure. In order to cure the very serious disease, Deng Xiao-Ping's taking method is to break the "iron bowl" and to develop the economy (to fill up the energies $\varphi(X) = U_f^{-2}$ and $\varphi(S_X) = R_f^{-2}$ of both wood (X) and water (S_X) at the same time, i.e., strengthen social structure and social benefit), and to allow a few people to get rich (to rush down the energy $\varphi(X_K) = \eta_f$ of soil (X_K), abate the coordinated ability). The idea is, at the same time, to use both "Strong inhibition of the same time, support the weak" and "Virtual disease is to fill his mother", if both S_X and X fall virtual diseases and at the same time, X_K befalls real disease.

The intervention method can be to maintain the balance of mathematical complex system because only two mathematical virtual disease subsystems are treated, by using Theorems 3.2 and 3.4, such that there is none side effects for all other subsystems. And the intervention method can also be to enhance the capability of intervention reaction and self-protection because the method of using intervention reaction and self-protection makes the ρ_3 and ρ_1 greater and near to 1. The state $\rho_3 = \rho_1 = 1$ is the best state of the steady multilateral system. On the way, it almost have none mathematical intervening resistance problem since any mathematical intervening method is possible good for some large ρ_3 and ρ_1.

CONCLUSIONS

This work shows how to treat the mathematical diseases (real or virtual) of a mathematical complex system in image mathematics and three methods are presented.

If only one subsystem falls ill, mainly the intervening method should be to intervene it indirectly for case: $\rho_2 = \rho_1\rho_3$, $\rho_3 = \rho_1$ and $\rho_1 \geq \rho_0$, according to the intervening principle of "Real disease is to rush down his son but virtual disease is to fill his mother". The intervention method directly can be used in case $\rho_2 = \rho_1\rho_3$, $\rho_3 = \rho_1$ and $\rho_1 < \rho_0$ but should be used as little as possible.

If two subsystems with the loving relation encounter sick, the intervening method should be intervene them directly according to the intervening principle of "Real disease is to rush down his son but virtual disease is to fill his mother".

If two subsystems with the killing relation encounter sick, the intervening method should be intervene them directly also according to the intervening principle of "Strong inhibition of the same time, support the weak".

Other properties, such as balanced, orderly nature of Wu-Xing, mathematical forecast, and so on, will be discussed in the next articles.

ACKNOWLEDGMENTS

This article has been repeatedly invited as reports, such as Shanxi University, Xuchang College, and so on. The work was supported by Specialized Research Fund for the Doctoral Program of Higher Education of Ministry of Education of China (Grant No. 200802691021).

REFERENCES

1. Y. S. Zhang, "Theory of Multilateral Matrices," Chinese Stat. Press, Beijing, 1993.
2. Y. S. Zhang, "Theory of Multilateral Systems," 2007. http://www.mlmatrix.com
3. Y. S. Zhang, "Mathematical Reasoning of Treatment Principle Based on 'Yin Yang Wu Xing' Theory in Traditional Chinese Medicine," Chinese Medicine, Vol. 2, No. 1, 2011, pp. 6-15. doi:10.4236/cm.2011.21002
4. Y. S. Zhang, "Mathematical Reasoning of Treatment Principle Based on 'Yin Yang Wu Xing' Theory in Traditional Chinese Medicine (II)," Chinese Medicine, Vol. 2, No. 4, 2011, pp. 158-170. doi:10.4236/cm.2011.24026
5. Y. S. Zhang, "Mathematical Reasoning of Treatment Principle Based on the Stable Logic Analysis Model of Complex Systems," Intelligent Control and Automation, Vol. 3, No. 1, 2012, pp. 6-15. doi:10.4236/ica.2012.31001

6. Y. S. Zhang, "Mathematical Reasoning of Economic Intevening Principle Based on 'Yin Yang Wu Xing' Theory in Traditional Chinese Economic (I)," Modern Economics, Vol. 3, No. 2, 2012, pp.

7. Y. S. Zhang, S. S. Mao, C. Z. Zhan and Z. G. Zheng, "Stable Structure of the Logic Model with Two Causal Effects," Chinese Journal of Applied Probability and Statistics, Vol. 21, No. 4, 2005, pp. 366-374.

8. C. Luo, X. D. Wang and Y. S. Zhang, "Orthogonality and Independence—New Thinking of Dealing with Complex Systems Series Three," Journal of Shanghai Institute of Technology (Natural Science), Vol. 10, No. 4, 2010, pp. 271-277.

9. C. Luo, X. P. Chen and Y. S. Zhang, "The Turning Point Analysis of Finance Time Series," Chinese Journal of Applied Probability and Statistics, Vol. 26, No. 4, 2010, pp. 437-442.

10. Y. S. Zhang, X. Q. Zhang and S. Y. Li, "SAS Language Guide and Application," Shanxi People's Press, Taiyuan, 2011.

11. Y. S. Zhang and S. S. Mao, "The Origin and Development Philosophy Theory of Statistics," Statistical Research, Vol. 12, 2004, pp. 52-59.

12. N. Q. Feng, Y. H. Qiu, F. Wang, Y. S. Zhang and S. Q. Yin, "A Logic Analysis Model about Complex System's Stability: Enlightenment from Nature," Lecture Notes in Computer Science, Vol. 3644, 2005, pp. 828-838. doi:10.1007/11538059_86

13. N. Q. Feng, Y. H. Qiu, Y. S. Zhang, F. Wang and Y. He, "A Intelligent Inference Model about Complex System's Stability: Inspiration from Nature," International Journal of Intelligent Technology, Vol. 1, 2005, pp. 1-6.

14. N. Q. Feng, Y. H. Qiu, Y. S. Zhang, C. Z. Zhan and Z. G. Zheng, "A Logic Analysis Model of Stability of Complex System Based on Ecology," Computer Science, Vol. 33, No. 7, 2006, pp. 213-216.

15. C. Y. Pan, X. P. Chen, Y. S. Zhang and S. S. Mao, "Logical Model of Five-Element Theory in Chinese Traditional Medicine," Journal of Chinese Modern Traditional Chinese Medicine, Vol. 4, No. 3, 2008, pp. 193- 196.

16. X. P. Chen, W. J. Zhu, C. Y. Pan and Y. S. Zhang, "Multilateral System," Journal of Systems Science, Vol. 17, No. 1, 2009, pp. 55-57.

17. C. Luo and Y. S. Zhang, "Framework Definition and Partition Theorems Dealing with Complex Systems: One of the Series of New Thinking," Journal of Shanghai Institute of Technology (Natural Science), Vol. 10, No. 2, 2010, pp. 109-114.

18. C. Luo and Y. S. Zhang, "Framework and Orthogonal Arrays: The New Thinking of Dealing with Complex Systems Series Two," Journal of Shanghai Institute of Technology (Natural Science), Vol. 10, No. 3, 2010, pp. 159-163.

19. J. Y. Liao, J. J. Zhang and Y. S. Zhang, "Robust Parameter Design on Launching an Object to Goal," Mathematics in Practice and Theory, Vol. 40, No. 24, 2010, pp. 126-132.

20. Y. S. Zhang, S. Q. Pang, Z. M. Jiao and W. Z. Zhao, "Group Partition and Systems of Orthogonal Idempotents," Linear Algebra and Its Applications, Vol. 278, No. 1-3, 1998, pp. 249-262. doi:10.1016/S0024-3795(97)10095-7

21. J. L. Zhao and Y. S. Zhang, "The Characteristic Description of Idempotent Orthogonal Class System," Advances in Matrix Theory and Its Applications, Proceedings of the 8th International Conference on Matrix and its applications, Taiyuan, Vol. 1, No. 1, 16-18 July 2008, pp. 445-448.

22. X. P. Chen, C. Y. Pan and Y. S. Zhang, "Partitioning the Multivariate Function Space into Symmetrical Classes," Mathematics in Practice and Theory, Vol. 39, No. 2, 2009, pp. 167-173.

23. C. Y. Pan, X. P. Chen and Y. S. Zhang, "Construct Systems of Orthogonal Idempotents," Journal of East China University (Natural Science), Vol. 141, No. 5, 2008, pp. 51-58.

24. C. Y. Pan, H. N. Ma, X. P. Chen and Y. S. Zhang, "Proof Procedure of Some Theories in Statistical Analysis of Global Symmetry," Journal of East China Normal University (Natural Science), Vol. 142, No. 5, 2009, pp. 127-137.

25. X. Q. Zhang, Y. S. Zhang and S. S. Mao, "Statistical Analysis of 2-Level Orthogonal Satursted Designs: The Procedure of Searching Zero Effects," Journal of East China Normal University (Natural Science), Vol. 24, No. 1, 2007, pp. 51-59.

26. Y. S. Zhang, Y. Q. Lu and S. Q. Pang, "Orthogonal Arrays Obtained by Orthogonal Decomposition of Projection Matrices," Statistica Sinica, Vol. 9, No. 2, 1999, pp. 595-604.

27. Y. S. Zhang, S. Q. Pang and Y. P. Wang, "Orthogonal Arrays Obtained by Generalized Hadamard Product," Discrete Mathematics, Vol. 238, No. 1-3, 2001, pp. 151- 170.doi:10.1016/S0012-365X(00)00421-0

28. Y. S. Zhang, L. Duan, Y. Q. Lu and Z. G. Zheng, "Construction of Generalized Hadamard Matrices $D(r^m(r + 1), r^m(r + 1); p)$," Journal of Statistical Planning, Vol. 104, 2002, pp. 239-258. doi:10.1016/S0378-3758(01)00249-X

29. Y. S. Zhang, "Data Analysis and Construction of Orthogonal Arrays," East China Normal University, Shanghai, 2006.

30. Y. S. Zhang, "Orthogonal Arrays Obtained by Repeating- Column Difference Matrices," Discrete Mathematics, Vol. 307, No. 2, 2007, pp. 246-261. doi:10.1016/j.disc.2006.06.029

31. X. D. Wang, Y. C. Tang, X. P. Chen and Y. S. Zhang, "Design of Experiment in Global Sensitivity Analysis Based on ANOVA High-Dimensional Model Representation," Communications in Statistics—Simulation and Computation, Vol. 39, No. 6, 2010, pp. 1183-1195. doi:10.1080/03610918.2010.484122

32. X. D. Wang, Y. C. Tang and Y. S. Zhang, "Orthogonal Arrays for the Estimation of Global Sensitivity Indices Based on ANOVA High-Dimensional Model Representation," Communications in Statistics—Simulation and Computation, Vol. 40, No. 9, 2011, pp. 1324-1341. doi:10.1080/03610918.2011.575500

33. J. T. Tian, Y. S. Zhang, Z. Q. Zhang, C. Y. Pan and Y. Y. Gan, "The Comparison and Application of Balanced Block Orthogonal Arrays and Orthogonal Arrays," Journal of Mathematics in Practice and Theory, Vol. 39, No. 22, 2009, pp. 59-67.

34. C. Luo and C. Y. Pan, "Method of Exhaustion to Search Orthogonal Balanced Block Designs," Chinese Journal of Applied Probability and Statistics, Vol. 27, No. 1, 2011, pp. 1-13.

35. Y. S. Zhang, W. G. Li, S. S. Mao and Z. G. Zheng, "Orthogonal Arrays Obtained by Generalized Difference Matrices with g Levels," Science China Mathematics, Vol. 54, No. 1, 2011, pp. 133-143. doi:10.1007/s11425-010-4144-y

36. Lao-tzu, "Tao Te Ching," In: S. Mitchel, Transl., 2010. http://acc6.its.brooklyn.cuny.edu

37. M.-J. Cheng, "Lao-Tzu, My Words Are Very Easy to Understand: Lectures on the Tao Teh Ching," North Atlantic Books, Richmond, 1981.

38. Research Center for Chinese and Foreign Celebrities and Developing Center of Chinese Culture Resources, "Chinese Philosophy Encyclopedia," Shanghai People Press, Shanghai, 1994.

CITATION

1. Y. Zhang and W. Shao, "Image Mathematics—Mathematical Intervening Principle Based on "Yin Yang Wu Xing" Theory in Traditional Chinese Mathematics (I)," Applied Mathematics, Vol. 3 No. 6, 2012, pp. 617-636. doi:10.4236/am.2012.36096.

First Review of Articles on Rhotrix Theory since Its Inception

A. Mohammed and M. Balarabe

Department of Mathematics, Ahmadu Bello University, Zaria, Nigeria

ABSTRACT

This paper presents an up-to-date review of the developments made in the field of rhotrix theory for a decade, starting from the year 2003, when the concept of rhotrix was introduced, up to the end of 2013. Over forty articles on rhotrix theory have been published in journals since its inception, indicating the need for a first review.

INTRODUCTION

In the year 2003, a relatively new paradigm of science, now known as rhotrix theory was initiated by Ajibade [1] , as an extension of ideas, on matrix-tertions and matrix-noitrets proposed by Atanassov and Shannon [2] . Since the publication of the article titled as "the concept of rhotrix for mathematical enrichment" in [1], many researchers have shown interest in the improvement of the theories and applications of rhotrices for the past one decade.

In the literature of rhotrix theory, starting from 2003, over forty articles have been published, thereby requiring the need for a first review. Before going further, it is pertinent to mention that two methods for multiplication of rhotrices having the same size are currently available in literature. The first one is "the heart based method for rhotrix multiplication" defined in [1], where the initial algebra and analysis of rhotrices were presented. The second alternative method is the row-column based method for rhotrix multiplication proposed by Sani [3] [4], in an attempt to answer the question of "finding a transformation for conversion rhotrix to matrix and vice versa" posed by Ajibade in the concluding section of his article. However, each method provides enabling environment to explore the usefulness of rhotrices as tools for carrying out mathematical research.

The objective of this article is to give a comprehensive literature survey of all published articles on rhotrix theory, since the introduction of the concept in 2003, up to the end of 2013. To achieve this, we classify all the over fourty articles in the literature of rhotrix theory into two classes. We term one class of the articles in the literature of rhotrix theory as commutative rhotrix theory, while the other class as non-commutative rhotrix theory. The reason behind this classification is due to the fact that, contributory author(s) of a single article on rhotrix theory adopted either Ajibade's heart-based method for multiplication of rhotrices or Sani's row-column method for multiplication of rhotrices in carrying out the work.

The choice of the two class names: commutative rhotrix theory and non-commutative rhotrix theory arise, respectively, from the commutative property inherent with the heart-based method for rhotrix multiplication, and the non-commutative property associated with row-column based method for rhotrix multiplication.

In line with this, articles on rhotrix theory can be broadly categorized according to the method of rhotrix multiplication used in presenting the work as follows:

1. Commutative rhotrix theory, i.e. Ajibade's article and all other articles using the Ajibade's heart-based method for rhotrix multiplication.

2. Non-commutative rhotrix theory, i.e. singularly authored articles by Sani and all other articles using Sani's row-column based method for rhotrix multiplication.

This survey paper contains three other sections after the introductory section. Section 2 presents the survey of developments in rhotrix theory. Section 3 analyzes these developments and then Section 4 presents the conclusion.

SURVEY OF DEVELOPMENTS ON RHOTRIX THEORY

This section presents a review of developments on rhotrix theory in a systematize form, starting with the review of commutative rhotrix theory in Subsection 2.1 and then followed by the review of non-commutative rhotrix theory in Subsection 2.2.

Class of Commutative Rhotrix Theory

Table 1 illustrates the title list of all journal articles that used Ajibade's heart based method for rhotrix multiplication, available in the literature of rhotrix theory, starting from 2003 and to the end of 2013. Thus, articles in Table 1 belong to the class of commutative rhotrix theory. Now, we start a systemic review of these works in Table 1 as follows:

In [1] Ajibade introduced the concept of rhotrix of size three as

$$\hat{R}(3) = \left\{ \left\langle \begin{array}{ccc} & a & \\ b & c & c \\ & e & \end{array} \right\rangle : a, b, c, d, e \in \Re \right\}$$

Where $h(R) = c$ is called the heart of any rhotrix $R \in R(3)$. The operations of addition, scalar multiplication and multiplication ($\circ$) are defined for rhotrices of size three in [1]. These rhotrix operations defined for rhotrix set of size three in [1] were thereafter, extended to rhotrix set of size n in the Ph.D. thesis of Mohammed [5], and recorded as follows: let

$$\hat{R}(n) = \left\{ \left\langle \begin{array}{ccccccccc} & & & & r_1 & & & & \\ & & & r_2 & r_3 & r_4 & & & \\ & & \cdots & \cdots & \cdots & \cdots & \cdots & & \\ & \cdots & \cdots & \cdots & \cdots & \cdots & \cdots & \cdots & \\ r_{\left[\frac{t+1}{2}\right]-n\backslash 2} & \cdots & \cdots & r_{\left[\frac{t+1}{2}\right]} & \cdots & \cdots & \cdots & r_{\left[\frac{t+1}{2}\right]+n\backslash 2} \\ & \cdots & \cdots & \cdots & \cdots & \cdots & \cdots & & \\ & & \cdots & \cdots & \cdots & \cdots & \cdots & & \\ & & r_{t-3} & r_{t-2} & r_{t-1} & & & \\ & & & r_t & & & & \end{array} \right\rangle : r_1, \cdots, r_t \in \Re \right\} \tag{1}$$

Table 1: List of titles in journals published from 2003 to 2013 that belong to the class of commutative rhotrix theory

S/no.	Title
1	A note on the rhotrix system of equations
2	A note on rhotrix exponent rule and its applications to special series and polynomial equation defined over rhotrices
3	A remark on the classifications of rhotrices as abstract structures
4	Algebraic properties of singleton, coiled and modulo rhotrices
5	Certain field of fractions
6	Certain quadratic extensions
7	Enrichment exercises through extension to rhotrices
8	Generalization and algorithmatization of heart based method for multiplication of rhotrices
9	Note on certain field of fractions
10	Note on rhotrices and the construction of finite fields
11	On construction of rhomtrees as graphical representation of rhotrices
12	On the structure of rhotrix
13	On the linear system over rhotrices
14	Rhotrices and the construction of finite fields
15	Rhotrix polynomials and polynomial rhotrices
16	Rhotrix sets and rhotrix spaces category
17	Rhotrix topological spaces
18	The concept of rhotrix in mathematical enrichment
19	The concept of heart oriented rhotrix multiplication

Be the set of all real rhotrices of size n, where, $n \in 2Z^+ + 1, t = \dfrac{1}{2}(n^2+1), n/2$

is the integer value obtained on division of n by 2, and $h(R) = r_{\left\{\frac{t+1}{2}\right\}}$ is the

heart of any rhotrix R in $\hat{R}(n)$. Let A(n) and B(n) be any two rhotrices in $\hat{R}(n)$ and scalar $\alpha \in \Re$. Then

$A(n) + B(n)$

$$
= \left\langle
\begin{array}{ccccccc}
 & & & a_1 + b_1 & & & \\
 & & a_2 + b_2 & a_3 + b_3 & a_4 + b_4 & & \\
 & \vdots & \vdots & \vdots & \vdots & \vdots & \\
a_{\left\{\frac{t+1}{2}\right\}-n\backslash2} + b_{\left\{\frac{t+1}{2}\right\}-n\backslash2} & \cdots & \cdots & h(A)h(B) & \cdots & \cdots & a_{\left\{\frac{t+1}{2}\right\}+n\backslash2} + b_{\left\{\frac{t+1}{2}\right\}+n\backslash2} \\
 & \vdots & \vdots & \vdots & \vdots & \vdots & \\
 & & a_{t-3}+b_{t-3} & a_{t-2}+a_{t-2} & a_{t-1}+b_{t-1} & & \\
 & & & a_t + b_t & & &
\end{array}
\right\rangle \quad (2)
$$

$$
\alpha A(n) = \left\langle
\begin{array}{ccccccc}
 & & & \alpha a_1 & & & \\
 & & \alpha a_2 & \alpha a_3 & \alpha a_4 & & \\
 & \vdots & \vdots & \vdots & \vdots & \vdots & \\
\alpha a_{\left\{\frac{t+1}{2}\right\}-n\backslash2} & \cdots & \cdots & \alpha a_{\left\{\frac{t+1}{2}\right\}} & \cdots & \cdots & \alpha a_{\left\{\frac{t+1}{2}\right\}+n\backslash2} \\
 & \vdots & \vdots & \vdots & \vdots & \vdots & \\
 & & a_{t-3} & a_{t-2} & a_{t-1} & & \\
 & & & a_t & & &
\end{array}
\right\rangle \quad (3)
$$

$A(n) \circ B(n) =$

$$
\left\langle
\begin{array}{ccccccc}
 & & & a_1 h(B)+b_1 h(A) & & & \\
 & & a_2 h(B)+b_2 h(A) & a_3 h(B)+b_3 h(A) & b_4 h(B)+b_4 h(A) & & \\
 & \vdots & \vdots & \vdots & \vdots & \vdots & \\
a_{\left\{\frac{t+1}{2}\right\}-n\backslash2} h(B)+b_{\left\{\frac{t+1}{2}\right\}-n\backslash2} h(A) & \cdots & \cdots & h(A)h(B) & \cdots & \cdots & a_{\left\{\frac{t+1}{2}\right\}+n\backslash2} h(B)+b_{\left\{\frac{t+1}{2}\right\}+n\backslash2} h(A) \\
 & \vdots & \vdots & \vdots & \vdots & \vdots & \\
 & & a_{t-3} h(B)+b_{t-3} h(A) & a_{t-2} h(B)+a_{t-2} h(A) & a_{t-1} h(B)+b_{t-1} h(A) & & \\
 & & & a_t h(B)+b_t h(A) & & &
\end{array}
\right\rangle, \quad (4)
$$

Where $h(A) = a_{\left(\frac{t+1}{2}\right)}$ and $h(B) = b_{\left(\frac{t+1}{2}\right)}$ are the hearts of rhotrices A(n) and B(n) respectively.

The extended rhotrix multiplication (4) was named in [5] as "Ajibade's heart-based method for multiplication of rhotrices". The rhotrix operations defined in [1] was adopted by [6] to present various classifications of rhotrices and their expressions as abstract structures of groups, semigroups, monoids, rings and Boolean algebras. The theorem for rhotrix exponent rule was first proposed without proof in [6], thereafter, [7] established and characterized the theorem for rhotrix exponent rule and extended the result to systemization of expressing special series and polynomial equations over rhotrices.

A remark on classifications of rhotrices as abstract structures was proposed by [8] over rhotrices. In the work, rhotrix ring was characterized; rhotrix integral domain and rhotrix field were constructed with certain conditions. Construction of certain field of fractions over rhotrices was presented by [9] as an extension to [8]. It was made known in [10] that the rhotrix field in [8] [9] holds only if the set of all hearty rhotrices of size three given in [6] is used as the underlying set.

The generalization of Ajibade's heart based method for rhotrix multiplication in [5] was algorithmatized for computing machines by [11]. A simplification of rhotrix expression generalization in [5] was presented by [12]. Construction and analysis of metric topological spaces using rhotrix set as the underlying set were considered in [13].

The concept of tree in graph theory was extended to rhotrix theory by [14] through their introduction of rhomtrees of order $m = \frac{1}{2}(n^2 + 1)$ as graphical representation of rhotrices of size n, where $n \in 2Z^+ + 1$. It was shown in their work that these rhomtrees have connection to known real world models such as topology of computing network, methane compound and certain product of sets.

In [15], the algebraic properties of singleton, coiled and modulo rhotrices were presented. Investigations of various constructions of finite fields over rhotrices were carried out in both [16] [17]. The cardinality of these finite fields was calculated through concrete examples. A study of the structure of rhotrices having entries from the set of integers modulo P and their properties was conducted by [18]. The rhotrix quadratic polynomial presented as part of a note on rhotrix exponent rule and its applications in [7] was given certain extensions by [19]. Rhotrix polynomial and its extension to construction of rhotrix polynomial ring was proposed in [20]. An investigation of rhotrix sets and rhotrix spaces categorized over numbers in real and complex fields was presented by [21]. A system of linear equations arising from the rhotrix equation $A \circ X = C$ was investigated in [22] and the conditions for their solvability were determined in the article. A note on rhotrix system of equations was presented by [23] as an extension to earlier work considered in [22]. The system of rhotrix equations was solved simultaneously.

Class of Non-Commutative Rhotrix Theory

Table 2 illustrates the title list of all journal articles that used row-column based method for rhotrix multiplication, available in the literature of rhotrix theory from 2003 to 2013. Thus, articles in Table 2 belong to the class of non-commutative rhotrix theory. Now, we start a systemic review of works in Table 2 as follows:

Sani [3] proposed (5) as an alternative method for multiplication of rhotrices of size three as an attempt to answer the question of "how can one convert a rhotrix to matrix and then vice versa", posed in the concluding section in [1].

$$R \circ Q = \left\langle \begin{matrix} & a & \\ b & h(R) & d \\ & e & \end{matrix} \right\rangle \circ \left\langle \begin{matrix} & f & \\ g & h(Q) & i \\ & j & \end{matrix} \right\rangle = \left\langle \begin{matrix} & af+bi & \\ ag+bj & h(R)h(Q) & ef+di \\ & dg+ej & \end{matrix} \right\rangle \tag{5}$$

This multiplication was later generalized by [4] to multiplication of rhotrices of size n as:

$$R(n) \circ S(n) = \left\langle a_{i,j}, b_{i,j} \right\rangle \circ \left\langle c_{k,l}, d_{k,l} \right\rangle = \left\langle \sum_{i,j=1}^{w} \left(a_{i,j}, b_{i,j} \right), \sum_{k,l=1}^{w-1} \left(c_{k,l}, d_{k,l} \right) \right\rangle, \tag{6}$$

Table 2: List of titles published from 2003 to 2013 that belong to the class of non-commutative rhotrix theory

S/no.	Titles
1	A determinant method for solving rhotrix system of eqn.
2	A note on relationship between invertible rhotrices and associated invertible matrices
3	Adjacent rhotrix of a complete, simple and undirected graph
4	Adjoint of a rhotrix and its basic properties
5	Algorithm design for row-column multiplication of n-dimensional rhotrices
6	An alternative method for multiplication of rhotrices
7	An example of linear mappings: extension to rhotrices
8	Cayley-Hamilton theorem in rhotrix
9	Conversion of a rhotrix to a coupled matrix
10	Hilbert matrix and its relationship with a special rhotrix
11	On inner product space and bilinear forms over rhotrices
12	On involutory and Pascal rhotrices
13	On the construction of involutory rhotrices.
14	Parallel multiplication of rhotrices using systolic array architecture
15	Rhotrix multiplication on two-dimensional process grid topologies
16	Rhotrices and elementary row operations
17	Rhotrix linear transformation
18	Rhotrix vector spaces
19	Row-wise representation of arbitrary rhotrix
20	Solution of two coupled matrices
21	The Cayley-Hamilton theorem for rhotrices
22	The equation $R_n(X) = b$, over rhotrices
23	The row-column multiplication of high dimensional rhotrices

Where $w = \dfrac{1}{2}(n+1)$ and $n \in 2Z^+ + 1$.

This (6) was presented in [5] as extended row-column based method for rhotrix multiplication.

In [24], a presentation of the concept of Hilbert rhotrix and its relationship with well-known Hilbert matrix was done. A special rhotrix termed as "Hilbert rhotrix" of size 5 was shown as a couple of two Hilbert matrices of sizes 3×3 and 2×2.

A method of converting rhotrix to a special form of matrix called "coupled matrix" was given in [25]. This was achieved through rotating the rhotrix R of size $n \in 2Z^{+} + 1$ at an angle of $45°$ in anti-clockwise direction, which result into a special form of matrix with missing values. For example, a rhotrix R of size 5 can express as a coupled matrix through half transpose as follows:

$$R(5)^{T/2} = \left\langle \begin{matrix} & & a_{11} & & \\ & a_{21} & c_{11} & a_{12} & \\ a_{31} & c_{21} & a_{22} & c_{12} & a_{13} \\ & a_{32} & c_{22} & a_{23} & \\ & & a_{33} & & \end{matrix} \right\rangle^{T/2} = \begin{bmatrix} a_{11} & & a_{12} & & a_{13} \\ & c_{11} & & c_{12} & \\ a_{21} & & a_{22} & & a_{23} \\ & c_{21} & & c_{22} & \\ a_{31} & & a_{32} & & a_{33} \end{bmatrix},$$

(7)

Where T/2 indicates a rotation through $45°$ in anti-clockwise direction. The special matrix in (7) is a coupling of 3×3 matrix with a 2×2 matrix, hence, the name "coupled matrix". Thus, in general,

$$R_n^{T/2} = \left\langle a_{ij}, c_{kl} \right\rangle^{T/2} = \left[a_{ij}, c_{kl} \right] = [Ac](n)$$

(8)

This is a rhotrix R(n) expressed as a coupled matrix of dimension n, coupling a $w \times w$ matrix with a $(w-1) \times (w-1)$ matrix, where $w = \frac{1}{2}(n+1)$.

Two coupled matrices $[Ac]_n$ and $[Bd]_n$ can be multiplied together by simply filling the missing spaces with zeros, after the multiplication, we removed the zero in other to have the result in filled coupled matrix form. The following is a very useful result recorded from Sani [25].

Theorem

If a coupled matrix $[Ac]_n$ is completed with zeros, then its determinants is the product of the determinants of the matrices $[A]_{w \times w}$ and $[C]_{(w-1) \times (w-1)}$, where $w = \dfrac{1}{2}(n+1)$.

This result on coupled matrix is very significant because it can be applied to solve problems involving two different systems of linear equations simultaneously, where one is a $t \times t$ system, $AX = b$ while the other is a $(t-1) \times (t-1)$ system $cY = d$.

Sani [26] presented the solution of two coupled matrices by extending the idea of a coupled matrix in [25] to a general case involving $m \times n$ and $(m-1) \times (m-1)$ matrices.

A one-sided system of the form $R_n(X) = b$, where R_n is an n-dimensional rhotrix, X the unknown n-dimensional rhotrix vector and b the right hand side rhotrix vector was presented by [27]. The necessary and sufficient conditions for the solvability of the system of an n-dimensional rhotrix equation $R_n \langle x^{ni} \rangle = \langle b^{nj} \rangle$ were discussed. Furthermore, the eigenvalues and the corresponding eigenvectors problems were solved.

The rhotrix addition and scalar multiplication defined in [1] was expressed in form of coupled matrices in [4]. The ideas were used by [28] to generalize and characterize the rhotrix vector space of size 3 initiated in [1] to rhotrix vector space of size n, through expression of rhotrices as coupled matrices.

Following this, [29] presented the concept of linear mapping to rhotrices and present its properties. It was shown in the work that the proposed method of converting a rhotrix to a "coupled matrix" (8) as defined in [25] is also a linear mapping.

In [30] an algorithm design for Sani's row-column based method for rhotrix multiplication (6) was proposed. As an extension to [3] [4] [25], various method of representing an arbitrary rhotrix was identified by [31]. One of the methods is the row-wise method, observed in the article to be flexible in analyzing rhotrices for mathematical enrichment. The flexibility of the representation has paved way for two formulae, one for row-column based method for arbitrary rhotrix multiplication and the other for heart-based method for arbitrary rhotrix multiplication.

The Cayley-Hamilton theorem for matrix is one of the well-known results in linear algebra. In 2012, the equivalence of this result was considered for rhotrix Cayley-Hamilton theorem in both [32] [33]. A note on relationship between invertible rhotrices and associated invertible matrices was proposed by [34]. A study of adjoint of a rhotrix and its basic properties was presented by [35]. The concept of inner product and bilinear forms over real rhotrices was considered in [36]. A determinant method for solving rhotrix system of linear equations was presented by [37].

It is well known that an involutory matrix is a matrix that is its own inverse. Such matrices are of great im- portance in matrix theory and algebraic cryptography. In [38] a method for constructing involutory rhotrices and their properties was given. Thereafter, an extension to [38] was given by [39] through the development of some theorems on involution in the context of rhotrices. Also, the description of Pascal rhotrices and their related properties was also considered. In [40], the theory of graph was extended to consider adjacent rhotrix of a complete, simple and undirected graph. A consideration of parallel multiplication of rhotrices using systolic array architecture was presented by [41]. Thereafter, a rhotrix multiplication on two-dimensional process grid topologies was carried out by [42]. The concept of rhotrix linear transformation with a number of theorems was presented by [43]. An investigation of rhotrices and its elementary row operations was carried out in [44].

Table 3: Number and percentage of articles on rhotrix theory per rhotrix theory class

S/no.	Category	Papers	Percentage (%)
1	Class of commutative rhotrix theory	19	45.24
2	Class of non-commutative rhotrix theory	23	54.46
Total		42	100

ANALYSIS

In this section, we present two tables for the analysis of articles in the literature review of rhotrix theory. In Table 3, we specify the number and percentage of articles on rhotrix theory from 2003 to the end of 2013 per rhotrix theory class.

The remarkable aspect of this literature review of articles on rhotrix theory is that authors following the class of commutative rhotrix theory enjoy the commutative property associated with the heart based method for rhotrix multiplication. For this reason, a number of abstract structures such as rhotrix groups, rhotrix semigroups, rhotrix rings, rhotrix Boolean algebra, rhotrix topological spaces, rhotrix metric spaces, rhotrix graphical trees called rhomtrees were developed. Furthermore, rhotrix finite fields, rhotrix exponent rule and their applications to special series, polynomial equations and polynomial rings over rhotrices were developed.

On the other hand, the contributory authors working on non-commutative rhotrix theory focus their researches majorly on extending the properties of matrices to rhotrices. Their inspirations came from the works of Sani [25] [26] in his papers on conversion of rhotrix to a special matrix termed as coupled matrix. These articles made several authors to study analogous properties of matrices to rhotrices.

Now, it is also pertinent for us to mention here that authors use the same symbol "∘" to denote both heart based rhotrix multiplication and row-column based rhotrix multiplication in their research papers. That could confuse readers as per which of the multiplication method was intended, particularly, when an algebraic structure is denoted as a pair. So to ensure clarity, it would be better for interested authors to use the symbol "∘" to denote heart based rhotrix multiplication and the symbol "•" to denote row-column based rhotrix multiplication in the future works.

In over all, we can say that from 2003 to 2013, the class of non-commutative rhotrix theory has more than 9% of articles in the literature of rhotrix theory than the class of commutative rhotrix theory.

CONCLUSIONS

In conclusion, we have presented a survey of articles on rhotrix theory starting from the year 2003 when the concept was initiated up to 2013. We have also classified the articles on rhotrix theory into two classes as commutative rhotrix theory and non-commutative rhotrix theory. It was shown in our analysis that the class of non- commutative rhotrix theory possessed 54.46% of articles in the literature of rhotrix theory while the class of commutative rhotrix theory possessed 45.24% of the articles.

ACKNOWLEDGMENTS

We wish to thank the unknown reviewers for their helpful suggestions. We also wish to thank Ahmadu Bello University, Zaria, Nigeria for funding this relatively new area of research.

REFERENCES

1. Ajibade, A.O. (2003) the Concept of Rhotrix in Mathematical Enrichment. International Journal of Mathematical Education in Science and Technology, 34, 175-179. http://dx.doi.org/10.1080/0020739021000053828
2. Atanassov, K.T. and Shannon, A.G. (1998) Matrix-Tertions and Matrix-Noitrets: Exercises in Mathematical Enrichment. International Journal of Mathematical Education in Science and Technology, 29, 898-903.
3. Sani, B. (2004) an Alternative Method for Multiplication of Rhotrices. International Journal of Mathematical Education in Science and Technology, 35, 777-781. http://dx.doi.org/10.1080/00207390410001716577
4. Sani, B. (2007) the Row-Column Multiplication of Higher Dimensional Rhotrices. International Journal of Mathematical Education in Science and Technology, 38, 657-662. http://dx.doi.org/10.1080/00207390601035245
5. Mohammed, A. (2011) Theoretical Development and Applications of Rhotrices. Ph.D. Thesis, Ahmadu Bello University, Zaria.
6. Mohammed, A. (2007) Enrichment Exercises through Extension to Rhotrices. International Journal of Mathematical Education in Science and Technology, 38, 131-136. http://dx.doi.org/10.1080/00207390600838490

7. Mohammed, A. (2007) a Note on Rhotrix Exponent Rule and Its Applications to Special Series and Polynomial Equations Defined over Rhotrices. Notes on Number Theory and Discrete Mathematics, 13, 1-15.

8. Mohammed, A. (2009) A Remark on the Classifications of Rhotrices as Abstract Structures. International Journal of Physical Sciences, 4, 496-499.

9. Usaini, S. and Tudunkaya, S.M. (2011) Certain Field of Fractions. Global Journal of Science Frontier Research, 11, 4-8.

10. Usaini, S. and Tudunkaya, S.M. (2012) Note on Certain Field of Fractions. Global Journal of Science Frontier Research, 12, 74-81.

11. Mohammed, A., Ezugwu, E.A. and Sani, B. (2011) On Generalization and Algorithmatization of Heart-Based Method for Multiplication of Rhotrices. International Journal of Computer Information Systems, 2, 46-49.

12. Absalom, E.A., Junaidu, S.B. and Sani, B. (2011) the Concept of Heart-Oriented Rhotrix Multiplication. Global Journal of Science Frontier, 11, 35-46.

13. Mohammed and Tijjani (2011) Rhotrix Topological Spaces. International Journal of Advances in Science and Technology, 3.

14. Mohammed, A. and Sani, B. (2011) On Construction of Rhomtrees as Graphical Representation of Rhotrices. Notes on Number Theory and Discrete Mathematics, 17, 21-29.

15. Tudunkaya and Makanjuola (2010) Algebraic Properties of Singleton, Coiled and Modulo Rhotrices. African Journal of Mathematics and Computer Sciences.

16. Tudunkaya, S.M. and Makanjuola, S.O. (2010) Rhotrices and the Construction of Finite Fields. Bulletin of Pure and Applied Sciences, 29e, 225-229.

17. Usaini, S. and Tudunkaya, S.M. (2011) Note on Rhotrices and the Construction of Finite Fields. Bulletin of Pure and Applied Sciences, 30e, 53-59.

18. Tudunkaya, S.M. and Makanjuola, S.O. (2012) On the Structure of Rhotrices. National Association of Mathematical Physics, 21, 271-280.

19. Tudunkaya, S.M. and Makanjuola, S.O. (2012) Certain Quadratic Extensions. National Association of Mathematical Physics, 21, 271-280.

20. Tudunkaya, S.M. (2013) Rhotrix Polynomial and Polynomial Rhotrices. Pure and Applied Mathematics Journal, 2, 38-41. http://dx.doi.org/10.11648/j.pamj.20130201.16

21. Mohammed, A. and Tella, Y. (2012) Rhotrix Sets and Rhotrix Spaces Category. International Journal of Mathematics and Computational Methods in Science and Technology, 2, 21-25.

22. Aminu, A. (2009) On the Linear System over Rhotrices. Notes on Number Theory and Discrete Mathematics, 15, 7-12.

23. Aminu, A. (2012) A Note on the Rhotrix System of Equation. Journal of the Nigerian Association of Mathematical Physics, 21, 289-296.

24. Kaurangini, M.L. and Sani, B. (2007) Hilbert Matrix and Its Relationship with a Special Rhotrix. Journal of the Mathematical Association of Nigeria, 34, 101-106.

25. Sani, B. (2008) Conversion of a Rhotrix to a Coupled Matrix. International Journal of Mathematical Education in Science and Technology, 39, 244-249. http://dx.doi.org/10.1080/00207390701500197

26. Sani, B. (2009) Solution of Two Coupled Matrices. Journal of the Mathematical Association of Nigeria, 32, 53-57.
27. Aminu, A. (2010) the Equation Rnx = b over Rhotrices. International Journal of Mathematical Education in Science and Technology, 41, 98-105. http://dx.doi.org/10.1080/00207390903189187
28. Aminu, A. (2010) Rhotrix Vector Spaces. International Journal of Mathematical Education in Science and Technology, 41, 531-578. http://dx.doi.org/10.1080/00207390903398408
29. Aminu, A. (2010) an Example of Linear Mappings: Extension to Rhotrices. International Journal of Mathematical Education in Science and Technology, 41, 691-698. http://dx.doi.org/10.1080/00207391003605213
30. Absalom, E.E., Ajibade, A.O. and Sahalu, J.B. (2011) Algorithm Design for Row-Column Multiplication of N-Di- mensional Rhotrices. Global Journal of Computer Science and Technology, 11, 22-30.
31. Chinedu, M.P. (2012) Row-Wise Representation of Arbitrary Rhotrix. Notes on Number Theory and Discrete Mathematics, 18, 1-27.
32. Sharma, P.L. and Kanwar, R.K. (2012) the Cayley-Hamilton Theorem for Rhotrices. International Journal of Mathematics and Analysis, 4, 171-178.
33. Aminu, A. (2012) Cayley-Hamilton Theorem in Rhotrices. National Association of Mathematical Physics, 20, 289- 296.
34. Sharma, P.L. and Kanwar, R.K. (2011) A Note on Relationship between Invertible Rhotrices and Associated Invertible Matrices. Bulletin of Pure and Applied Sciences: Mathematics and Statistics, 30e, 333-339.
35. Sharma, P.L. and Kanwar, R.K. (2012) Adjoint of a Rhotrix and Its Basic Properties. International Journal of Mathematical Sciences, 11, 337-343.
36. Sharma, P.L. and Kanwar, R.K. (2012) On Inner Product Spaces and Bilinear Forms of Rhotrices. Bulletin of Pure and Applied Sciences, 31e, 109-118.
37. Aminu, A. (2012) A Determinant Method for Solving Rhotrix System of Equation. Journal of the Nigerian Association of Mathematical Physics, 21, 281-288.
38. Usaini, S. (2012) On the Construction of Involutory Rhotrices. International Journal of Mathematical Education in Science and Technology, 43, 510-515. http://dx.doi.org/10.1080/0020739X.2011.599875
39. Sharma, P.L. and Kanwar, R.K. (2013) On Involutory and Pascal Rhotrices. International Journal of Mathematical Sciences and Engineering Applications, 7, 133-146.
40. Aminu, A. and Michael, O. (2013) Adjacent Rhotrix of a Complete, Simple and Undirected Graph. National Association of Mathematical Physics, 25, 267-274.
41. Absalom, E.A., Abdullahi, M., Ibrahim, K., Mohammed, A. and Junaidu, S.B. (2011) Parallel Multiplication of Rhotrices Using Systolic Array Architecture. International Journal of Computer Information Systems, 2, 68-73.
42. Absalom, E.E. and Sahalu, J.B. (2012) Rhotrix Multiplication of Two-Dimensional Process Grid Topologies. International Journal of Grid and High Performance Computing, 4, 21-36. http://dx.doi.org/10.4018/jghpc.2012010102

43. Mohammed, A., Balarabe, M. and Imam, A.T. (2012) Rhotrix Linear Transformation. Advances in Linear Algebra & Matrix Theory, 2, 43-47. http://dx.doi.org/10.4236/alamt.2012.24007
44. Usaini, S. (2012) Rhotrices and Elementary Row Operation. Journal of the Nigerian Association of Mathematical Physics, 20, 37-42.

CITATION

1. Mohammed, A. and Balarabe, M. (2014) First Review of Articles on Rhotrix Theory since Its Inception. Advances in Linear Algebra & Matrix Theory, 4, 216-224. doi: 10.4236/alamt.2014.44020.

Monty Hall Problem and the Principle of Equal Probability in Measurement Theory

Shiro Ishikawa

Department of Mathematics, Faculty of Science and Technology, Keio University, Yokohama, Japan

6

ABSTRACT

In this paper, we study the principle of equal probability (i.e., unless we have sufficient reason to regard one possible case as more probable than another, we treat them as equally probable) in measurement theory (i.e., the theory of quantum mechanical world view), which is characterized as the linguistic turn of quantum mechanics with the Copenhagen interpretation. This turn from physics to language does not only realize the remarkable extension of quantum mechanics but also establish the method of science. Our study will be executed in the easy example of the Monty Hall problem. Although our argument is simple, we believe that it is worth pointing out the fact that the principle of equal probability can be, for the first time, clarified in measurement theory (based on the dualism) and not the conventional statistics (based on Kolmogorov's probability theory).

INTRODUCTION

Monty Hall Problem

The Monty Hall problem is well-known and elementary. Also it is famous as the problem in which even great mathematician P. Erdös made a mistake (cf. [1]). The Monty Hall problem is as follows:

Problem 1 [Monty Hall problem 1]: You are on a game show and you are given the choice of three doors. Behind one door is a car, and behind the other two are goats. You choose, say, door 1, and the host, who knows where the car is, opens another door, behind which is a goat. For example, the host says that (b) the door 3 has a goat.

And further, He now gives you the choice of sticking with door 1 or switching to door 2? What should you do?

In the framework of measurement theory [2-12], we shall present two answers of this problem in Sections 3.1 and 4.2. Although this problem seems elementary, we assert that the complete understanding of the Monty Hall problem can not be acquired within Kolmogorov's probability theory [13] but measurement theory (based on the dualism).

Overview: Measurement Theory

As emphasized in refs. [7, 8], measurement theory (or in short, MT) is, by a linguistic turn of quantum mechanics (cf. Figure 1: ③ later), constructed as the scientific theory formulated in a certain C^*-algebra A (i.e., a norm closed subalgebra in the operator algebra B(H) composed of all bounded operators on a Hilbert space H, cf. [14,15]). MT is composed of two theories (i.e., pure measurement theory (or, in short, PMT] and statistical measurement theory (or, in short, SMT). That is, it has the following structure:

(A) MT (measurement theory)

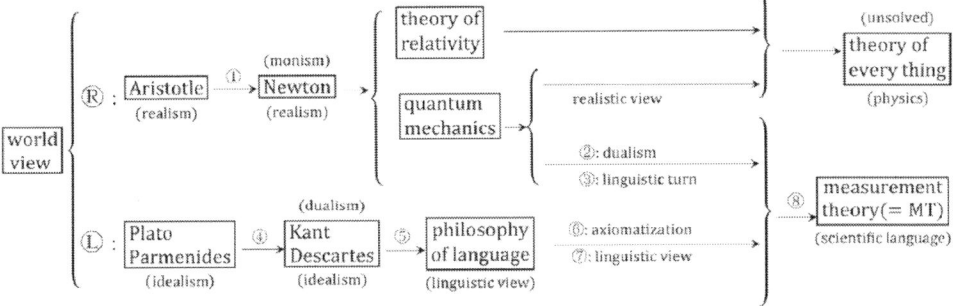

Figure 1: The development of the world views from our standing point. For the explanation of (①-⑧), see [8, 10].

$$\begin{cases} (A_1):[PMT] \\ =\big[(pure)\,measurement\big]+[causality] \\ \qquad \big(Axiom^P 1\big) \qquad\qquad (Axiom2) \\[2em] (A_2):[SMT] \\ =\big[(statistical)\,measurement\big]+[causality] \\ \qquad \big(Axiom^S 1\big) \qquad\qquad (Axiom2) \end{cases}$$

where Axiom 2 is common in PMT and SMT. For completeness, note that measurement theory (A) (i.e., (A$_1$) and (A$_2$)) is not physics but a kind of language based on "the (quantum) mechanical world view". As seen in [9], note that MT gives a foundation to statistics. That is, roughly speaking (B) it may be understandable to consider that PMT and

SMT is related to Fisher's statistics and Bayesian statistics respectively.

Also, for the position of MT in science, see Figure 1, which was precisely explained in [8, 10].

When A = B$_c$(H), the C*-algebra composed of all compact operators on a Hilbert space H, the (A) is called quantum measurement theory (or, quantum system theory), which can be regarded as the linguistic aspect of quantum mechanics. Also, when A is commutative (that is, when A is characterized by $C_0(\Omega)$, the C*-algebra composed of all continuous complex-valued functions vanishing at infinity on a locally compact Hausdorff space Ω (cf. [16])), the (A) is called classical measurement theory. Thus, we have the following classification:

$$\text{(C)} \quad \text{MT} \begin{cases} \text{quantum MT} \left(\text{when } A = B_c(H) \right) \\ \text{classical MT} \left(\text{when } A = C_0(\Omega) \right) \end{cases}$$

The purpose of this paper is to clarify the Monty Hall problem in the classical PMT and classical SMT.

CLASSICAL MEASUREMENT THEORY (AXIOMS AND INTERPRETATION)

Mathematical Preparations

Since our concern is the Monty Hall problem, we devote ourselves to classical MT in (C). Throughout this paper, we assume that Ω is a compact Hausdorff space. Thus, we can put $C_0(\Omega) = C(\Omega)$, which is defined by a Banach space (or precisely, a commutative C*-algebra) composed of all continuous complex-valued functions on a compact Hausdorff space Ω, where its norm $\|f\|_{C(\Omega)}$ is defined by $\max_{\omega \in \Omega} |f(\omega)|$. Let $C(\Omega)^*$ be the dual Banach space of $C(\Omega)$. That is, $C(\Omega)^* = \{\rho \mid \rho \text{ is a continuous linear functional on } C(\Omega)\}$, and the norm $\|\rho\|_{C(\Omega)^*}$ is defined by $\sup\{|\rho(f)| : f \in C(\Omega) \text{ such that } \|f\|_{C(\Omega)} \leq 1\}$.

The bi-linear functional $\rho(f)$ is also denoted by $_{C(\Omega)^*}\langle \rho, f \rangle_{C(\Omega)}$, or in short $\langle \rho, f \rangle$.

Define the mixed state $\rho\left(\in C(\Omega)^*\right)$ such that $\|f\|_{C(\Omega)^*} = 1$ and $\rho(f) \geq 0$ for all $f \in C(\Omega)$ such that $f \geq 0$. And put

$$S^m\left(C(\Omega)^*\right) = \left\{\rho \in C(\Omega)^* \mid \rho \text{ is a mixed state}\right\}$$

Also, for each $\omega \in \Omega$, define the pure state $\delta_\omega\left(\in S^m\left(C(\Omega)^*\right)\right)$ such that $_{C(\Omega)^*}\langle \delta_\omega, f\rangle_{C(\Omega)} = f(\omega)\left(\forall f \in C(\Omega)\right)$. And put

$$S^p\left(C(\Omega)^*\right) = \left\{\delta_\omega \in C(\Omega)^* \mid \delta_\omega \text{ is a pure state}\right\}$$

which is called a state space. Note, by the Riesz theorem (cf. [16]), that $C(\Omega)^* = M(\Omega) \equiv \left\{\rho \mid \rho \text{ is a signed measure on } \Omega\right\}$ and $S^m\left(C(\Omega)^* \right) = M_{+1}^m(\Omega) \equiv \left\{\rho \mid \rho \text{ is a measure on } \Omega \text{ such that } \rho(\Omega) = 1\right\}$. Also, it is clear that $S^p\left(C(\Omega)^* \right) = \left\{\rho_{\omega_0} \mid \rho_{\omega_0} \text{ is a point measure at } \omega_0 \in \Omega\right\}$, where

$$\int_\Omega f(\omega)\delta_{\omega_0}(d\omega) = f(\omega_0)\left(\forall f \in C(\Omega)\right)$$

This implies that the state space $S^p\left(C(\Omega)^*\right)$ can be also identified with Ω (called a spectrum space or simply, spectrum) such as

$$S^p\left(C(\Omega)^*\right) \ni \delta_\omega \leftrightarrow \omega \in \Omega$$

(state space)   (spectrum)   (1)

Also, note that $C(\Omega)$ is unital, i.e., it has the identity I (or precisely, $I_{C(\Omega)}$), since we assume that Ω is compact.

According to the noted idea (cf. [17]) in quantum mechanics, an observable $O := (X, B_X, F)$ in $C(\Omega)$ is defined as follows:

(D_1) [Field] X is a set, $B_X (\subseteq 2^X$, the power set of $X)$ is a field of X, that is, "$\Xi_1, \Xi_2 \in B_X \Rightarrow \Xi_1 \cup \Xi_2 \in B_X$", $\Xi \in B_X \Rightarrow X \setminus \Xi \in B_X$".

(D_2) [Additivity] F is a mapping from B_X to $C(\Omega)$ satisfying: 1): for every $\Xi \in B_X$, $F(\Xi)$ is a non-negative element in $C(\Omega)$ such that $0 \leq F(\Xi) \leq I$, 2): $F(\Phi) = 0$ and $F(X) = I$, where 0 and I is the 0-element and the identity in $C(\Omega)$ respectively. 3): for any $\Xi_1, \Xi_2 \in B_X$ such that $\Xi_1 \cap \Xi_2 = \Phi$, it holds that $F(\Xi_1 \cup \Xi_2) = F(\Xi_1) + F(\Xi_2)$.

For the more precise argument (such as countably additivity, etc.), see [7, 9].

Classical PMT in (A_1)

In this section we shall explain classical PMT in (A_1).

With any system S, a commutative C^*-algebra $C(\Omega)$ can be associated in which the measurement theory (A) of that system can be formulated. A state of the system S is represented by an element $\delta_\omega (\in S^P (C(\Omega)^*))$ and an observable is represented by an observable $O := (X, B_X, F)$ in $C(\Omega)$. Also, the measurement of the observable O for the system S with the state δ_ω is denoted by $M_{C(\Omega)} (O, S_{[\delta_\omega]})$ (or more precisely $M_{C(\Omega)} (O := (X, B_X, F), S_{[\delta_\omega]}))$. An observer can obtain a measured value $x (\in X)$ by the measurement $M_{C(\Omega)} (O, S_{[\delta_\omega]})$.

The Axiom^P 1 presented below is a kind of mathematical generaliza-tion of Born's probabilistic interpretation of quantum mechanics. And thus, it is a statement without reality.

Axiom^P 1 [Measurement]: The probability that a measured value $x(\in X)$ obtained by the measurement $M_{C(\Omega)}\left(O := (X, B_x, F), S_{[\delta_\omega]}\right)$ be-longs to a set $\Xi(\in B_x)$ is given by $[F(\Xi)](\omega_0)$.

Next, we explain Axiom 2 in (A). Let $(T, \leq)$ be a tree, i.e., a partial ordered set such that "$t_1 \leq t_3$ and $t_2 \leq t_3$" implies "$t_1 \leq t_2$ or $t_2 \leq t_1$" In this paper, we assume that T is finite. Also, assume that there exists an element $t_0 \in T$, called the root of T, such that $t_0 \leq t(\forall t \in T)$ holds. Put $T_\leq^2 = \left\{(t_1, t_2) \in T^2 \mid t_1 \leq t_2\right\}$. The family $\left\{\Phi_{t_1, t_2} : C(\Omega_{t_2}) \to C(\Omega_{t_1})\right\}_{(t_1, t_2) \in T_\leq^2}$ is called a causal relation (due to the Heisenberg picture), if it satisfies the following conditions (E_1) and (E_2).

(E_1) With each $t \in T$, a C^*-algebra $C(\Omega_t)$ is associated.

(E_2) For every $(t_1, t_2) \in T_\leq^2$, a Markov operator $\Phi_{t_1, t_2} : C(\Omega_{t_2}) \to C(\Omega_{t_1})$ is defined (i.e., $\Phi_{t_1, t_2} \geq 0$, $\Phi_{t_1, t_2}\left(I_{C(\Omega_{t_2})}\right) = I_{C(\Omega_{t_1})}$). And it satisfies that $\Phi_{t_2, t_3} = \Phi_{t_1, t_3}$ holds for any $(t_1, t_2)(t_2, t_3) \in T_\leq^2$.

The family of dual operators

$$\left\{\Phi_{t_1, t_2}^* : S^m\left(C(\Omega_{t_1})^*\right) \to S^m\left(C(\Omega_{t_2})^*\right)\right\}_{(t_1, t_2) \in T_\leq^2}$$

is called a dual causal relation (due to the Schrödinger picture). When

$$\Phi_{t_1, t_2}^*\left(S^p\left(C(\Omega_{t_1})^*\right)\right) \subseteq S^p\left(C(\Omega_{t_2})^*\right)$$

holds for any $(t_1, t_2) \in T_{\le}^2$, the causal relation is said to be deterministic.

Here, Axiom 2 in the measurement theory (A) is presented as follows:

Axiom 2 [Causality]: The causality is represented by a causal relation $\left\{ \Phi_{t_1, t_2} : C\left(\Omega_{t_2}\right) \to C\left(\Omega_{t_1}\right) \right\}_{(t_1, t_2) \in T_{\le}^2}$

For the further argument (i.e., the W^*-algebraic formulation) of measurement theory, see Appendix in [7].

Classical SMT in (A_2)

It is usual to consider that we do not know the state δ_{ω_0} when we take a measurement $M_{C(\Omega)}\left(O, S_{[\delta_\omega]}\right)$. That is because we usually take a measurement $M_{C(\Omega)}\left(O, S_{[\delta_{\omega_0}]}\right)$ in order to know the state δ_{ω_0}. Thus, when we want to emphasize that we do not know the state δ_{ω_0}, $M_{C(\Omega)}\left(O, S_{[\delta_{\omega_0}]}\right)$ is denoted by $M_{C(\Omega)}\left(O, S_{[*]}\right)$. Also, when we know the distribution $v_0 \left(\in M_{+1}^m(\Omega) = S^m\left(C(\Omega)^*\right) \right)$ of the unknown state δ_{ω_0}, the $M_{C(\Omega)}\left(O, S_{[\delta_{\omega_0}]}\right)$ is denoted by $M_{C(\Omega)}\left(O, S_{[*]}(\{v_0\})\right)$.

The Axiom^S 1 presented below is a kind of mathematical generalization of Axiom^P 1.

Axiom^S 1 [Statistical measurement]: The probability that a measured value $x(\in X)$ obtained by the measurement $M_{C(\Omega)}\left(O := (X, B_x, F), S_{[*]}(\{v_0\})\right)$ belongs to a set $\Xi(\in F)$ is given by

$$v_0\left(F(\Xi)\right)\left(=_{C(\Omega)^*} F(\Xi)_{C(\Omega)}\right).$$

Remark 1: Note that two statistical measurements $M_{C(\Omega)}\left(O, S_{[\delta_{\omega_1}]}(\{v_0\})\right)$ and $M_{C(\Omega)}\left(O, S_{[\delta_{\omega_2}]}(\{v_0\})\right)$ can not be distinguished before measurements. In this sense, we consider that, even if $\omega_1 \neq \omega_2$, we can assume that

$$M_{C(\Omega)}\left(O, S_{[\delta_{\omega_1}]}(\{v_0\})\right) = M_{C(\Omega)}\left(O, S_{[*]}(\{v_0\})\right)$$

$$= M_{C(\Omega)}\left(O, S_{[\delta_{\omega_2}]}(\{v_0\})\right)$$

(2)

Linguistic Interpretation

Next, we have to answer how to use the above axioms as follows. That is, we present the following linguistic interpretation (F) [= $(F_1) - (F_3)$], which is characterized as a kind of linguistic turn of so-called Copenhagen interpretation (cf. [7, 8]). That is, we propose:

(F_1) Consider the dualism composed of "observer" and "system (= measuring object)". And therefore, "observer" and "system" must be absolutely separated.

(F_2) Only one measurement is permitted. And thus, the state after a measurement is meaningless since it can not be measured any longer. Also, the causality should be assumed only in the side of system, however, a state never moves. Thus, the Heisenberg picture should be adopted.

(F_3) Also, the observer does not have the space-time. Thus, the question: "When and where is a measured value obtained?" is out of measurement theory, and so on. This interpretation is, of course, common to both PMT and SMT.

Remark 2: Note that quantum mechanics has many interpretations (i.e., several Copenhagen interpretation, many worlds interpretation, statistical interpretation, etc.). On the other hand, we believe that the interpretation of measurement theory (A) is uniquely determined as in the above. This is our main reason to propose the linguistic interpretation of quantum mechanics. We believe that this uniqueness is essential to the justification of Heisenberg's uncertainty principle (cf. [10, 18]).

Preliminary Fundamental Theorems

We have the following two fundamental theorems in measurement theory:

Theorem 1: [Fisher's maximum likelihood method (cf. [9])]. Assume that a measured value obtained by a measurement $M_{C(\Omega)}\left(O := (X, B_x, F), S_{[*]}\right)$ belongs to $\Xi(\in B_x)$. Then, there is a reason to infer that the unknown state [*] is equal to δ_{ω_0}, where $\omega_0 (\in \Omega)$ is defined by

$$\left[F(\Xi)\right](\omega_0) = \max_{\omega \in \Omega}\left[F(\Xi)\right](\omega)$$

Theorem 2: [Bayes' method (cf. [9])]. Assume that a measured value obtained by a statistical measurement $M_{C(\Omega)}\left(O := (X, B_x, F), S_{[*]}(\{v_0\})\right)$ belongs to $\Xi(\in B_x)$.

Then, there is a reason to infer that the posterior state (i.e., the mixed state after the measurement) is equal to v_{post}, which is defined by

$$v_{post}(D) = \frac{\int_D \left[F(\Xi)\right](\omega)v_0(d\omega)}{\int_\Omega \left[F(\Xi)\right](\omega)v_0(d\omega)}$$

$(\forall D \in B_\Omega;$ Borel field$)$.

The above two theorems are, of course, the most fundamental in statistics. Thus, if we believe in Figure 1, we can answer to the following problem (cf. [4, 9]):

(G) What is statistics? Or, where is statistics in science? which is certainly the most essential problem in the philosophy of statistics.

THE FIRST ANSWER TO MONTY HALL PROBLEM

Fisher's Method (The First Answer)

In this section, we present the first answer to Problem 1 (Monty-Hall problem) in classical PMT. Put

$\Omega = \{\omega_1, \omega_2, \omega_3\}$ with the discrete topology. Assume that each state

$\delta_{\omega_m} \left(\in S^p \left(C(\Omega)^* \right) \right)$ means

$\delta_{\omega_m} \Leftrightarrow$ the state that the car is behind the door 1

$(m = 1, 2, 3)$

$\qquad\qquad\qquad\qquad\qquad\qquad\qquad\qquad\qquad\qquad$ (3)

Define the observable $O_1 \equiv \left(\{1, 2, 3\}, 2^{\{1, 2, 3\}}, F_1 \right)$ in $C(\Omega_t)$ such that

$$\left[F_1 \left(\{1\} \right) \right] (\omega_1) = 0.0, \quad \left[F_1 \left(\{2\} \right) \right] (\omega_1) = 0.5,$$

$$\left[F_1 \left(\{3\} \right) \right] (\omega_1) = 0.5, \quad \left[F_1 \left(\{1\} \right) \right] (\omega_2) = 0.0,$$

$$\left[F_1 \left(\{2\} \right) \right] (\omega_2) = 0.0, \quad \left[F_1 \left(\{3\} \right) \right] (\omega_2) = 1.0,$$

$$\left[F_1 \left(\{1\} \right) \right] (\omega_3) = 0.0, \quad \left[F_1 \left(\{2\} \right) \right] (\omega_3) = 1.0,$$

$$\left[F_1 \left(\{3\} \right) \right] (\omega_3) = 0.0,$$

$\qquad\qquad\qquad\qquad\qquad\qquad\qquad\qquad\qquad\qquad$ (4)

where it is also possible to assume that $F_1 (\{2\}) (\omega_1) = \alpha$, $F_1 (\{3\}) (\omega_1) =$ $= 1 - \alpha (0 < \alpha < 1)$. Thus we have a measurement $M_{C(\Omega)} \left(O_1, S_{[*]} \right)$, which should be regarded as the measurement theoretical representation of the measurement that you say "door 1". Here, we assume that

1. "measured value is obtained by the measurement $M_{C(\Omega)} \left(O_1, S_{[*]} \right) \Leftrightarrow$ The host says "Door 1 has a goat";

2. "measured value is obtained by the measurement $M_{C(\Omega)}\left(O_1, S_{[*]}\right) \Leftrightarrow$ The host says "Door 1 has a goat";

3. "measured value is obtained by the measurement $M_{C(\Omega)}\left(O_1, S_{[*]}\right) \Leftrightarrow$ The host says "Door 1 has a goat".

Recall that, in Problem 1, the host said "Door 3 has a goat". This implies that you get the measured value "3" by the measurement $M_{C(\Omega)}\left(O_1, S_{[*]}\right)$. Therefore, Theorem 1 (Fisher's maximum likelihood method) says that you should pick door number 2. That is because we see that

$$\left[F_1(\{3\})\right](\omega_2) = 1.0 = \max\{0.5, 1.0, 0.0\}$$
$$= \max\left\{\left[F_1(\{3\})\right](\omega_1), \left[F_1(\{3\})\right](\omega_2), \left[F_1(\{3\})\right](\omega_3)\right\},$$

$$(5)$$

and thus, there is a reason to infer that $[*] = \delta_{\omega_2}$. Thus, you should switch to door 2. This is the first answer to Problem 1 (the Monty-Hall problem 1).

Bayes' Method (Answer to Modified Monty Hall Problem 2)

In the sense mentioned in Remark 3 later, the following modified Monty Hall problem (Problem 2) is completely different from Problem 1 (the Monty Hall problem 1). However, it is worth examining Problem 2 for the better understanding of Problem 3 later.

Problem 2: [Modified Monty Hall problem 2]. Suppose you are on a game show, and you are given the choice of three doors (i.e., "number 1", "number 2", "number 3"). Behind one door is a car, behind the others, goats. You pick a door, say number 1. Then, the host, who set a car behind a certain door, says

($\#_1$) the car was set behind the door decided by the cast of the distorted dice. That is, the host set the car behind the k-th door (i.e., "number k") with probability p_k (or, weight such that $p_1 + p_2 + p_3 = 1, 0 \le p_1, p_2, p_3 \le 1$).

And further, the host says, for example (b) the door 3 has a goat.

He says to you, "Do you want to pick door number 2?" Is it to your advantage to switch your choice of doors?

In what follows we study this problem. Let Ω and O_1 be as in Section 3.1. Under the hypothesis ($\#_1$), define the mixed state $v_0 \left(\in M_{+1}^m (\Omega) \right)$ such that:

$$v_0 \left(\{ \omega_1 \} \right) = p_1, v_0 \left(\{ \omega_2 \} \right) = p_2, v_0 \left(\{ \omega_3 \} \right) = p_3 \tag{6}$$

Thus we have a statistical measurement

$M_{C(\Omega)} \left(O_1, S_{[*]} \left(\{ v_0 \} \right) \right)_{\parallel} \Leftrightarrow$. Note that

1. "measured value is obtained by the statistical measurement $M_{C(\Omega)} \left(O_1, S_{[*]} \left(\{ v_0 \} \right) \right) \Leftrightarrow$ The host says "Door 1 has a goat";

2. "measured value is obtained by the statistical measurement $M_{C(\Omega)} \left(O_1, S_{[*]} \left(\{ v_0 \} \right) \right) \Leftrightarrow$ The host says "Door 2 has a goat";

3. "measured value is obtained by the statistical measurement $M_{C(\Omega)} \left(O_1, S_{[*]} \left(\{ v_0 \} \right) \right) \Leftrightarrow$ The host says "Door 1 has a goat".

Here, assume that, by the statistical measurement $M_{C(\Omega)} \left(O_1, S_{[*]} \left(\{ v_0 \} \right) \right)$, you obtain a measured value 3which corresponds to the fact that the host said "Door 3 has a goat". Then, Theorem 2 (Bayes' theorem) says that the posterior state $v_{\text{post}} \left(\in M^m_{+1}(\Omega) \right)$ is given by

$$v_{\text{post}} = \frac{F_1\left(\{3\}\right) \times v_0}{\left\langle v_0, F_1\left(\{3\}\right) \right\rangle} \tag{7}$$

That is,

$$v_{\text{post}}\left(\{\omega_1\}\right) = \frac{\dfrac{p_1}{2}}{\dfrac{p_1}{2} + p_2}, \quad v_{\text{post}}\left(\{\omega_2\}\right) = \frac{p_2}{\dfrac{p_1}{2} + p_2},$$

$$v_{\text{post}}\left(\{\omega_3\}\right) = 0. \tag{8}$$

Particularly, we see that (H) if $p_1 = p_2 = p_3 = 1/3$, then it holds that $v_{\text{post}}\left(\{\omega_1\}\right) = \frac{1}{3}$, $v_{\text{post}}\left(\{\omega_2\}\right) = \frac{2}{3}$, $v_{\text{post}}\left(\{\omega_3\}\right) = 0$, and thus, you should pick Door 2.

Remark 3: The difference between Problem 1 and Problem 2 should be remarked. Since the $(\#_1)$ in Problem 2 is the information from the host to you, Problem 1 and Problem 2 are completely different. Although the above (H) may be generally regarded as the proper answer of the Monty Hall problem, we do not admit that the (H) is proper. That is, we consider that the (H) is not the second answer to the Monty Hall problem.

THE SECOND ANSWER TO MONTY HALL PROBLEM

In this section, we shall present the second answer. However, before it, we have to prepare the principle of equal probability (i.e., unless we have sufficient reason to regard one possible case as more probable than another, we treat them as equally probable). For completeness, note that measurement theory urges us to use only Axioms 1 and 2.

The Principle of Equal Probability

Put $\Omega = \{\omega_1, \omega_2, \omega_3, ..., \omega_n\}$ with the discrete topology. And consider any observable $O_1 \equiv (X, B_x, F_1)$ in $C(\Omega)$.

Define the bijection $\phi_1 : \Omega \to \Omega$ such that

$$\phi_1(\omega_j) = \begin{cases} \omega_{j+1} & (j \neq n) \\ \omega_1 & (j = n) \end{cases}$$

and define the observable $O_k \equiv (X, B_x, F_k)$ in $C(\Omega)$ such that

$$\left[F_k(\Xi) \right](\omega) = \left[F_1(\Xi) \right](\phi_{k-1}(\omega))$$
$$(\forall \omega \in \Omega, k = 1, 2, \cdots, n)$$

where $\phi_0(\omega) = \omega (\forall \in \Omega)$ and

$$\phi_k(\omega) = \phi_1(\phi_{k-1}(\omega))(\forall \omega \in \Omega, k = 1, 2, \cdots, n)$$

Let $p_k (k = 1, ..., n)$ be a non-negative real number such that $\Sigma_{k=1}^{n} p_k = 1$.

(I) For example, fix a state $\delta_{\omega_m} (m = 1, 2, ..., n)$. And, by the cast of the distorted dice, you choose an observable $O_k \equiv (X, B_x, F_k)$ with probability p_k. And further, you take a measurement

$$M_{C(\Omega)} \left(O_k := (X, B_X, F_k), S_{[\delta_{\omega_m}]} \right)$$

Here, we can easily see that the probability that a measured value obtained by the measurement (I) belongs to $\Xi (\in B_x)$ is given by

$$\sum_{k=1}^{n} p_k \left\langle F_k(\Xi), \delta_{\omega_m} \right\rangle \left(= \sum_{k=1}^{n} p_k \left[F_k(\Xi) \right](\omega_m) \right)$$

(9)

which is equal to $\left\langle F_1(\Xi), \sum_{k=1}^{n} p_k \delta_{\phi_{k-1}(\omega_m)} \right\rangle$. This implies that the measurement (I) is equivalent to a statistical measurement:

$$M_{C(\Omega)}\left(O_1 := (X, B_X, F_1), S_{\left[\delta_{\omega_m}\right]}\left(\left\{\sum_{k=1}^{n} p_k \delta_{\phi_{k-1}(\omega_m)}\right\}\right)\right)$$

Note that the (9) depends on the state δ_m. Thus, we can not calculate the (9) such as the (8).

However, if it holds that $p_k = \frac{1}{n}(k = 1, ..., n)$, we see that $\frac{1}{n}\sum_{k=1}^{n}\delta_{\phi_{k-1}(\omega_m)}$ is independent of the choice of the state δ_{ω_m}. Thus, putting $\frac{1}{n}\sum_{k=1}^{n}\delta_{\phi_{k-1}(\omega_m)} = v_e$, we see that the measurement (I) is equivalent to the statistical measurement $M_{C(\Omega)}\left(O_1, S_{\left[\delta_{\omega_m}\right]}(\{v_e\})\right)$, which is also equivalent to $M_{C(\Omega)}\left(O_1, S_{[*]}(\{v_e\})\right)$ (from the formula (2) in Remark 1).

Thus, under the above notation, we have the following theorem.

Theorem 3: [The principle of equal probability (i.e., the equal probability of selection)]. If $p_k = \frac{1}{n}(k = 1, ..., n)$, the measurement (I) is independent of the choice of the state δ_m. Hence, the (I) is equivalent to a statistical measurement

$$M_{C(\Omega)}\left(O_1 := (X, B_X, F_1), S_{[*]}(\{v_e\})\right)$$

It should be noted that the principle of equal probability is not "principle" but "theorem" in measurement theory.

Remark 4: This theorem was also discussed in [5, 6], where we missed the formula (2) in Remark 1. Thus, the argument in [5, 6] was too abstract. And thus, it might be regarded as ambiguous and vague. In fact, we must admit that the explanation in [5, 6] is not yet accepted generally. Therefore, we recommend readers to read [5, 6] after the understanding of the concrete explanation (I) in the linguistic interpretation (F). Also, note that Theorem 3 is independent of Axiom 2. And further, for the principle of equal (a priori) probabilities in equilibrium statistical mechanics, see [11], in which how to use measurement theory (and thus statistics) in statistical mechanics is explained.

The Second Answer to Monty Hall Problem (i.e., Modified Monty Hall Problem 3)

As an application of Theorem 3, we consider the following modified Monty-Hall problem:

Problem 3: [Modified Monty Hall problem 3]. Suppose you are on a game show, and you are given the choice of three doors (i.e., "number 1", "number 2", "number 3"). Behind one door is a car, behind the others, goats.

($\#_2$) You choose a door by the cast of the fair dice, i.e., with probability 1/3.

According to the rule ($\#_2$), you pick a door, say number 1, and the host, who knows where the car is, opens another door, behind which is a goat. For example, the host says that

(b) the door 3 has a goat.

He says to you, "Do you want to pick door number 2?" Is it to your advantage to switch your choice of doors?

[Answer]. Consider Ω and O_1 as in Section 3.1. Then, Theorem 3 says that the answer of Problem 3 is the same as the (H). Thus, you should pick the door 2.

Remark 5: The difference between the $(\#_1)$ in Problem 2 and the $(\#_2)$ in Problem 3 is clear in the dualism (F). The former is host's selection, but the latter is your selection (i.e., observer's selection). That is, in Problem 3, the information from host to you is only the (b). This situation is the same as that of Problem 1. In this sense, we think that Problems 1 and 3 are similar. That is, we can conclude that Problem 1 [resp. Problem 3] is the Monty Hall problem in PMT [resp. SMT]. Also, our recent report [19] will promote a better understanding of measurement theory.

CONCLUSIONS

In the conventional statistics based on Kolmogorov's probability theory, Problem 3 may be unconsciously confused with Problem 2. On the other hand, as mentioned in Remark 5, the difference between Problems 2 and 3 can be clearly described in measurement theory (based on the dualism (F)). This is the merit of measurement theory.

What we executed in this paper may be merely the translation from "ordinary language" to "scientific language", that is,

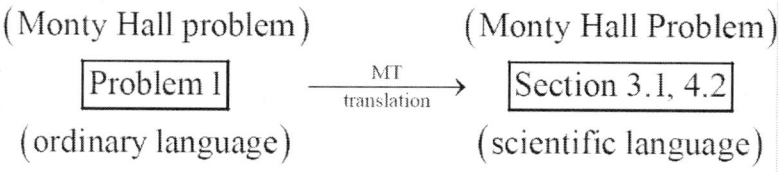

We believe that this translation is just "the mechanical world view" or "the method of science" (at least, science in the series L of Figure 1). That is, ordinary science (at least, its basic statements) should be described in terms of measurement theory. For example, for the translation of equilibrium statistical mechanics and the Zeno's paradoxes, see [11] and [12] respectively. Probably, we refrained from the publication of [12], if we were not sure of "MT = the method of science (or the form of scientific thinking)".

In this paper (as well as [9]), we showed one of advantages of the measurement theoretical foundation of statistics through the examination of the Monty Hall problem. Also, recall that measurement theory possesses a great power to answer to the problem (G). However, our methodology should be tested from various points of view, because the classic statistics methodology (based on Kolmogorov's probability theory) can be good applied in many fields. We hope that our approach will be examined from various view points.

REFERENCES

1. P. Hoffman, "The Man Who Loved Only Numbers, the Story of Paul Erdös and the Search for Mathematical Truth," Hyperion, New York, 1998,
2. S. Ishikawa, "Fuzzy Inferences by Algebraic Method," Fuzzy Sets and Systems, Vol. 87, No. 2, 1997, pp. 181-200. doi:10.1016/S0165-0114(96)00035-8
3. S. Ishikawa, "A Quantum Mechanical Approach to Fuzzy Theory," Fuzzy Sets and Systems, Vol. 90, No. 3, 1997, pp. 277-306. doi:10.1016/S0165-0114(96)00114-5
4. S. Ishikawa, "Statistics in Measurements," Fuzzy Sets and Systems, Vol. 116, No. 2, 2000, pp. 141-154. doi:10.1016/S0165-0114(98)00280-2
5. S. Ishikawa, "Mathematical Foundations of Measurement Theory," Keio University Press Inc., 2006. http://www.keio-up.co.jp/kup/mfomt/
6. S. Ishikawa, "Monty Hall Problem in Unintentional Random Measurements," Far East Journal of Dynamical Systems, Vol. 3, No. 2, 2009, pp. 165-181.
7. S. Ishikawa, "A New Interpretation of Quantum Mechanics," Journal of Quantum Information Science, Vol. 1, No. 2, 2011, pp. 35-42. doi:10.4236/jqis.2011.12005
8. S. Ishikawa, "Quantum Mechanics and the Philosophy of Language: Reconsideration of Traditional Philosophies," Journal of Quantum Information Science, Vol. 2, No. 1, 2012, pp. 2-9. doi:10.4236/jqis.2012.21002
9. S. Ishikawa, "A Measurement Theoretical Foundation of Statistics," Applied Mathematics, Vol. 3, No. 3, 2012, pp. 183-192.
10. S. Ishikawa, "The Linguistic Interpretation of Quantum Mechanics," 2012. http://arxiv.org/pdf/1204.3892.pdf
11. S. Ishikawa, "Ergodic Hypothesis and Equilibrium Statistical Mechanics in the Quantum Mechanical World View," World Journal of Mechanics, Vol. 2, No. 2, 2012, pp. 125-130.doi:10.4236/wjm.2012.22014
12. S. Ishikawa, "Zeno's Paradoxes in the Mechanical World View," 2012. http://arxiv.org/pdf/1205.1290.pdf
13. A. Kolmogorov, "Foundations of the Theory of Probability (Translation)," Chelsea Pub Co. Second Edition, New York, 1960.
14. G. J. Murphy, "C*-Algebras and Operator Theory," Academic Press, Boston, 1990.

15. J. von Neumann, "Mathematical Foundations of Quantum Mechanics," Springer Verlag, Berlin, 1932.
16. K. Yosida, "Functional Analysis," 6th Edition, SpringerVerlag, Berlin, 1980.
17. E. B. Davies, "Quantum Theory of Open Systems," Academic Press, London, 1976.
18. S. Ishikawa, "Uncertainty Relation in Simultaneous Measurements for Arbitrary Observables," Reports on Mathematical Physics, Vol. 29, No. 3, 1991, pp. 257-273.doi:10.1016/0034-4877(91)90046-P
19. S. Ishikawa, "What Is Statistics? The Answer by Quantum Language," arXiv:1207.0407v1 [physics.data-an], 2012. http://arxiv.org/abs/1207.0407v1

CITATION

1. S. Ishikawa, "Monty Hall Problem and the Principle of Equal Probability in Measurement Theory," Applied Mathematics, Vol. 3 No. 7, 2012, pp. 788-794. doi: 10.4236/am.2012.37117.

Contractions of Certain Lie Algebras in the Context of the DLF-Theory

Alexander Levichev and Oleg Sviderskiy
Sobolev Institute of Mathematics, Novosibirsk, Russia

ABSTRACT

Contractions of the Lie algebras d = u (2), f = u (1, 1) to the oscillator Lie algebra l are realized via the adjoint action of SU (2, 2) when d, l, f are viewed as subalgebras of su(2, 2). Here D, L, F are the corresponding (four-dimensional) real Lie groups endowed with bi-invariant metrics of Lorentzian signature. Similar contractions of (seven-dimensional) isometry Lie algebras iso(D), iso(F) to iso(L) are determined. The group SU (2, 2) acts on each of the D, L, F by conformal transformation which is a core feature of the DLF-theory. Also, d and f are contracted to T, S-abelian subalgebras, generating parallel translations, T, and proper conformal transformations, S (from the decomposition of su(2, 2) as a graded algebra T + Ω + S, where Ω is the extended Lorentz Lie algebra of dimension 7).

INTRODUCTION

As noticed by the first author (see [1, 2]), there are precisely three four-dimensional non-abelian Lie algebras that admit a non-degenerate invariant bilinear form of Lorentzian signature:

The oscillator Lie algebra l, defined by the following commutation relations in a certain basis l_1, l_2, l_3, l_4:

$$[l_2,l_3] = l_1, [l_2,l_4] = l_3, [l_3,l_4] = -l_2;$$

$$(1.1)$$

the Lie algebra d = u(2) generated by vectors X_0, X_1, X_2, X_3 with the following commutation relations:

$$[X_1,X_2] = 2X_3, [X_1,X_3] = -2X_2, [X_2,X_3] = 2X_1;$$

$$(1.2)$$

the Lie algebra f=u(1,1) generated by vectors H_0, H_1, H_2, H_3, satisfying the following commutation relations

$$[H_0,H_1] = 2H_2, [H_1,H_2] = -2H_0, [H_0,H_2] = -2H_1$$

$$(1.3)$$

Relations (1.2) are the same as the ones in [3].

Certain Lie groups (corresponding to the Lie algebras d, l, f), endowed with bi-invariant metrics of Lorentzian signature, provide so-called homogeneous solutions to Einstein's equations of General Relativity Theory, GRT. We denote these solutions by the corresponding capital letters. They have been studied in [4]. In GRT literature (see [5]), D is known as the perfect fluid, F is a tachyonic fluid, and that L is a very special case of plane waves. Namely, it is an isotropic electromagnetic field determined by a covariantly constant light-like vector (see [6]). Groups defined by the oscillator Lie algebra l are also studied in detail on pp. 409-414 of [7].

Relations (1.1), (1.2), and (1.3) have been used in [8, 9] where the basics of the DLF-approach have been presented (by A. L.). This DLF-theory can be briefly characterized as the LF-modification of Segal's Chronometry (the latter one is based on D, see [3, 9] and references therein).

It is known (see [9]) that these three space-times are conformally flat, and that in each case the isometry group of the corresponding Lorentzian manifold is of dimension 7. Each of d, l, f (as well as a four-dimensional abelian Lie algebra) can be realized as a subalgebra of the conformal algebra su(2, 2) and these imbeddings will be specified below. Conventions about su(2, 2) follow [3, p. 92]. Namely, a traceless four by four matrix m (with complex entries allowed) is in su(2,2) iff

$$m * s + sm = 0,$$

Where s is a diagonal matrix:

$$s = \begin{bmatrix} 1 & 0 & 0 & 0 \\ 0 & 1 & 0 & 0 \\ 0 & 0 & -1 & 0 \\ 0 & 0 & 0 & -1 \end{bmatrix}.$$

Also, commutation relations relative to a certain basis in su(2,2) are (3.1) of our Section 3.

The general subject of contractions (and deformations) of Lie algebras is important in physics (see, for example, Section IV.7 of [10]). It provides an explanation of one physical model being a limiting case of another one. In [10], it was the case of Newtonian world versus Minkowski space-time M. The findings of the present article (together with what has been already published by the first author) form the necessary base for investigation of similar relationships between space-times M, D, L, F.

The subject is also of interest from a pure mathematical point of view. Regarding contractions of Lie algebras, we follow [11].

LIE-THEORETICAL CONTRACTIONS OF d AND f

Namely, we use the name Lie-theoretical contraction for the method defined by [11, p.2, (3)], where the commutation relations of a contracted Lie algebra are given by

$$[x, y]' = \lim_{\delta x \to 0} U_q^{-1}\left(\left[U_q(x), U_q(y)\right]\right).$$

Here q=p+δx When q is not equal to p, U_q is a non-singular linear transformation of the original Lie algebra. When q = p, the inverse of the linear transformation U_p does not exist.

Theorem 2.1: There is a Lie-theoretical contraction of d=u (2) to the oscillator Lie algebra l.

Proof: Consider the following vectors in d:

$$\begin{cases} L_1 = -\alpha^2\left(X_0 + X_3\right), L_2 = \alpha X_2, \\ L_3 = \alpha X_1, L_4 = -(1/2)\left(X_0 - X_3\right). \end{cases}$$

One can verify that for any non-zero value of α, the algebra generated by L_1, L_2, L_3, L_4, is isomorphic to d, since the Equations (2.1) are uniquely solvable for X_i.

The commutation relations for L_1, L_2, L_3, L_4, are as follows:

$$\begin{cases} \left[L_2, L_3\right] = L_1 - 2\alpha^2 L_4, \left[L_2, L_4\right] = L_3, \left[L_3, L_4\right] = -L_2, \\ \left[L_1, L_2\right] = 2\alpha^2 L_3, \left[L_1, L_3\right] = -2\alpha^2 L_2. \end{cases}$$

It can be easily seen that as α goes to zero, the commutation relations become (1.1), that is, of the oscillator Lie algebra l.

Remark 2.1: Alternatively, we can choose a different set of vectors in d:

$$L_1 = \beta^2 \left(X_0 - X_3 \right), L_2 = \beta X_2, L_3 = \beta X_1, L_4 = \left(1/2 \right)\left(X_0 + X_3 \right).$$

It defines a Lie algebra isomorphic to d (when β is not zero). The commutation relations converge to those of the oscillator algebra when β goes to zero.

Theorem 2.2: There is a Lie-theoretical contraction of f=u (1, 1) to the oscillator Lie algebra l.

Proof: We apply similar contraction procedures to f as we did in the case of d. Let

$$L_1 = -\alpha^2 \left(H_0 + H_3 \right), L_2 = \alpha H_1, L_3 = \alpha H_2, L_4 = \left(1/2 \right)\left(H_0 - H_3 \right)$$

Then, as α goes to 0, the commutation relations for L_1, L_2, L_3, L_4, become (1.1), that is, of the oscillator algebra l. Clearly, for any nonzero α, the Lie algebra generated by L_1, L_2, L_3, L_4, is isomorphic to f.

Remark 2.2: As noticed in [12], there always exists a (trivial) Lie-theoretical contraction of any Lie algebra to an abelian algebra.

REALIZATION IN su (2, 2)

Our current goal is to realize the above-mentioned contractions of the Lie algebras d and f through the adjoint action of the group SU (2, 2) on its Lie algebra su(2, 2) of which all d, f and l are subalgebras. We will call such contractions the su(2, 2)-inner contractions of the corresponding Lie algebras.

More specifically, we will conjugate the generators of d and f with elements of the maximal abelian subgroup A of SU (2, 2), from the Iwasawa decomposition SU (2, 2) =KAN.

Remark 3.1: A. L. thanks David Vogan for helpful discussions on the subject.

Remark 3.2: It is important to bear in mind (see [8, 9]) that SU (2, 2) acts on each of D, L, F. Joint consideration of the three worlds is the key feature of the DLF-theory.

In [13, p.135], the generic element a of the two-dimensional maximal abelian subalgebra A of SU (2, 2) is chosen as

$$\begin{bmatrix} 0 & 0 & s & 0 \\ 0 & 0 & 0 & t \\ s & 0 & 0 & 0 \\ 0 & t & 0 & 0 \end{bmatrix},$$

Where s, t are real parameters. It thus can be written in terms of the generators L_{ij} of SU (2, 2) as

$$s\left(L_{-14} - L_{03}\right) + t\left(L_{-14} + L_{03}\right).$$

All fifteen matrices L_{ij} (where I, j=-1, 0,..., 4) forming a (standard) basis of su(2, 2) can be found in [14]. Relative to this basis, the su(2, 2) commutation relations are as follows

$$\left[L_{ij}, L_{jk}\right] = e_j L_{ik},$$

$$(3.1)$$

Where $e_{-1}=e_0=-1$, $e_1=\ldots=e_4=1$

The generic element of the corresponding group A can then be written as

$$a(s,t) = \exp \begin{bmatrix} 0 & 0 & s & 0 \\ 0 & 0 & 0 & t \\ s & 0 & 0 & 0 \\ 0 & t & 0 & 0 \end{bmatrix} = \begin{bmatrix} \cosh s & 0 & \sinh s & 0 \\ 0 & \cosh t & 0 & \sinh t \\ \sinh s & 0 & \cosh s & 0 \\ 0 & \sinh t & 0 & \cosh t \end{bmatrix}$$

Clearly,

$$a^{-1}(s,t) = \begin{bmatrix} \cosh s & 0 & -\sinh s & 0 \\ 0 & \cosh t & 0 & -\sinh t \\ -\sinh s & 0 & \cosh s & 0 \\ 0 & -\sinh t & 0 & \cosh t \end{bmatrix}.$$

The following imbedding of d=u (2) into su(2, 2) has been presented in [14, p.5262]:

$$X_0 = L_{-10}, X_1 = L_{14} - L_{23}, X_2 = L_{24} + L_{13}, X_3 = L_{34} - L_{12}. \tag{3.2}$$

Theorem 3.1: There exist su(2, 2)-inner contractions of d to oscillator subalgebras of su(2,2). Besides, there exist su(2, 2)-inner contractions of d to abelian subalgebras of su(2,2).

Proof: Choose the following basis for the Lie algebra d:

$$L_{-10} - L_{34} + L_{12}, L_{24} + L_{13}, L_{14} - L_{23}, L_{34} - L_{12},$$

and conjugate the corresponding matrices with a generic element a=a(s, t) of the group A. Direct calculation shows that

$$\begin{cases} a(\mathbf{L}_{-10} - \mathbf{L}_{34} + \mathbf{L}_{12})a^{-1} = \cosh 2t(\mathbf{L}_{-10} - \mathbf{L}_{34}) + \sinh 2t(\mathbf{L}_{-13} - \mathbf{L}_{04}) + \mathbf{L}_{12}, \\ a(\mathbf{L}_{24} + \mathbf{L}_{13})a^{-1} = \cosh(s-t)\mathbf{L}_{13} + \sinh(s-t)\mathbf{L}_{01} + \cosh(s+t)\mathbf{L}_{24} - \sinh(s+t)\mathbf{L}_{-12}, \\ a(\mathbf{L}_{-14} - \mathbf{L}_{23})a^{-1} = \cosh(s+t)\mathbf{L}_{14} - \sinh(s+t)\mathbf{L}_{-11} - \cosh(s-t)\mathbf{L}_{23} - \sinh(s-t)\mathbf{L}_{02}, \\ a(\mathbf{L}_{-10} + \mathbf{L}_{34} - \mathbf{L}_{12})a^{-1} = \cosh 2s(\mathbf{L}_{-10} + \mathbf{L}_{34}) - \sinh 2s(\mathbf{L}_{-13} + \mathbf{L}_{04}) - \mathbf{L}_{12}. \end{cases} \tag{3.3}$$

Set t = 0 in the above system and rewrite the equations as:

$$\begin{cases} a(s,0)(\mathbf{L}_{-10} - \mathbf{L}_{34} + \mathbf{L}_{12})a^{-1}(s,0) \\ = \mathbf{L}_{-10} - \mathbf{L}_{34} + \mathbf{L}_{12}, a(s,0)(\mathbf{L}_{24} + \mathbf{L}_{13})a^{-1}(s,0) = (e^s/2)(\mathbf{L}_{13} + \mathbf{L}_{01} + \mathbf{L}_{24} - \mathbf{L}_{-12}) \\ + (e^{-s}/2)(\mathbf{L}_{13} - \mathbf{L}_{01} + \mathbf{L}_{24} + \mathbf{L}_{-12}), a(s,0)(\mathbf{L}_{-14} - \mathbf{L}_{23})a^{-1}(s,0) = (e^s/2) \\ (\mathbf{L}_{14} - \mathbf{L}_{-11} - \mathbf{L}_{23} - \mathbf{L}_{02}) + (e^{-s}/2)(\mathbf{L}_{14} + \mathbf{L}_{-11} - \mathbf{L}_{23} + \mathbf{L}_{02}), \\ a(s,0)(\mathbf{L}_{-10} + \mathbf{L}_{34} - \mathbf{L}_{12})a^{-1}(s,0) = (e^{2s}/2)(\mathbf{L}_{-10} + \mathbf{L}_{34} - \mathbf{L}_{-13} - \mathbf{L}_{04}) \\ + (e^{-2s}/2)(\mathbf{L}_{-10} + \mathbf{L}_{34} + \mathbf{L}_{-13} + \mathbf{L}_{04}) - \mathbf{L}_{12}. \end{cases}$$

Now, introduce from the equations above.

$$l_1 = -e^{-2s}a(s,0)(\mathbf{L}_{-10} + \mathbf{L}_{34} - \mathbf{L}_{12})a^{-1}(s,0),$$

$$l_2 = e^{-s}a(s,0)(\mathbf{L}_{24} + \mathbf{L}_{13})a^{-1}(s,0), l_3 = e^{-s}a(s,0)(\mathbf{L}_{14} - \mathbf{L}_{23})a^{-1}(s,0),$$

$$l_4 = -(1/2)a(s,0)(\mathbf{L}_{-10} - \mathbf{L}_{34} + \mathbf{L}_{12})a^{-1}(s,0),$$

What we observe here is a special case of the contraction (2.1) with $\alpha = e^{-s}$.

As s goes to infinity, the commutation relations become (1.1), which means that the outcome of the contraction is a subalgebra of su(2, 2), isomorphic to the oscillator algebra I. Namely, basic matrices of new subalgebra are the limits of the matrices l_1, l_2, l_3, l_4.

$$\begin{cases} q_1 = -(1/2)(L_{-10} + L_{34} - L_{-13} - L_{04}), \\ q_2 = (1/2)(L_{13} + L_{01} + L_{24} - L_{-12}), \\ q_3 = (1/2)(L_{14} - L_{-11} - L_{23} - L_{02}), \\ q_4 = -(1/2)(L_{-10} - L_{34} + L_{12}), \end{cases}$$

$$(3.4)$$

Alternatively, we can set s = 0 in system (3.3), and after a similar procedure, we get a limiting Lie algebra spanned by

$$m_1 = (1/2)(L_{-10} - L_{34} + L_{-13} - L_{04}),$$

$$m_2 = (1/2)(L_{13} - L_{01} + L_{24} - L_{-12}),$$

$$m_3 = (1/2)(L_{14} - L_{-11} - L_{23} + L_{02}),$$

$$m_4 = (1/2)(L_{-10} + L_{34} - L_{12}).$$

It is also isomorphic to the oscillator Lie algebra, since the commutation relations are (1.1).

Finally, setting s = t in system (3.2), and repeating the procedure above, realizes a contraction of d to an abelian subalgebra of su(2, 2). The limiting algebra T (as s = t go to infinity) is generated by

$$T_0 = (1/2)(L_{-10} - L_{04}), T_1 = (1/2)(L_{-11} - L_{14}), T_2 = (1/2)(L_{-12} - L_{24}), T_3 = (1/2)(L_{-13} - L_{34}).$$

If we make s = t go to negative infinity, then the resulting abelian Lie algebra S is generated by

$$S_0 = (1/2)(L_{-10} + L_{04}), S_1 = (1/2)(L_{-11} + L_{14}), S_2 = (1/2)(L_{-12} + L_{24}), S_3 = (1/2)(L_{-13} + L_{34}).$$

This finishes the proof of Theorem 3.1.

Remark 3.3: The (above) two subalgebras are known (in that order) as the Lie algebra of translations and the Lie algebra of "proper conformal transformations" of the Minkowski space-time: see [15], where su(2, 2) is written as a graded algebra T + Ω + S with T, S being the two abelian algebras above, and Ω being the Lorentz Lie algebra extended by (infinitesimal) dilatations. The above choice of generators for T and for S has been made in Table V of [3].

In the remaining part of this section we arrange for similar procedures, as in the above, but for the case of f=u(1, 1). We realize it as a su(2, 2)-subalgebra by choosing the following basis:

$$H_0 = L_{-10} - L_{12}, H_1 = L_{01} + L_{-12}, H_2 = L_{02} - L_{-11}, H_3 = L_{34}. \tag{3.5}$$

This choice of the basis, with commutation relations (1.3), has been made in [8].

Theorem 3.2: There exists a su(2, 2)-inner contraction of this subalgebra f to a subalgebra of su(2,2) isomorphic to the oscillator Lie algebra l.

Proof: Choose one other basis for f:

$$H_0 + H_3, H_1, H_2, H_0 - H_3. \tag{3.6}$$

Conjugating these vectors with a(s, t) as in the proof of Theorem 3.1, we get the following:

$$
\begin{cases}
a\left(L_{-10} + L_{34} - L_{12}\right)a^{-1} = \cosh 2s\left(L_{-10} + L_{34}\right) - \sinh 2s\left(L_{-13} + L_{04}\right) - L_{12}, \\
a\left(L_{01} + L_{-12}\right)a^{-1} = \cosh(s-t)L_{01} + \sinh(s-t)L_{13} + \cosh(s+t)L_{-12} - \sinh(s+t)L_{24}, \\
a\left(L_{02} - L_{-11}\right)a^{-1} = \cosh(s-t)L_{02} + \sinh(s-t)L_{23} - \cosh(s+t)L_{-11} + \sinh(s+t)L_{14}, \\
a\left(L_{-10} - L_{34} - L_{12}\right)a^{-1} = \cosh 2t\left(L_{-10} - L_{34}\right) - \sinh 2t\left(L_{-13} - L_{04}\right) - L_{12}.
\end{cases} \tag{3.7}
$$

If we set t = 0 and perform a contraction analogous to that for d, we get in the limit as s goes to infinity, a subalgebra of su(2, 2) generated by the following set:

$$\begin{cases} V_1 = -(1/2)\big(L_{-10} + L_{34} - L_{-13} - L_{04}\big), \\ V_2 = (1/2)\big(L_{01} + L_{13} + L_{-12} - L_{24}\big), \\ V_3 = (1/2)\big(L_{02} + L_{23} - L_{-11} + L_{14}\big), \\ V_4 = -(1/2)\big(L_{-10} - L_{34} - L_{12}\big). \end{cases} \tag{3.8}$$

This subalgebra is isomorphic to the oscillator Lie algebra since the commutation relations are the same as (1.1).

CONTRACTIONS OF THE ISOMETRY LIE ALGEBRAS

As mentioned in the Introduction, the isometry groups of the Lorentzian manifolds, corresponding to the Lie algebras d, f, and l are of dimension 7. All three of these 7-dimensional Lie algebras, iso(D), iso(F), and iso(L) can be viewed as subalgebras of su(2, 2). The corresponding imbeddings are specified below (in Theorems 4.1, 4.2, 4.3), where $Z_{su(2,2)}$ (b) will denote the centralizer of a vector b in su(2, 2). Considerable part of observations in this section has been known before, and we indicate a few references (see below). However, bringing together different methods and applying them to each of our main objects of study (space-times D, L, F), make the content of the section to a new ingredient of the mathematical presentation of the DLF-theory.

According to [3], the following elements form a basis for k=iso (D):

$$\begin{cases} X_0 = L_{-10}, X_1 = L_{14} - L_{23}, X_2 = L_{24} - L_{-13}, X_3 = L_{34} - L_{12}, \\ Y_1 = L_{14} + L_{23}, Y_2 = L_{24} - L_{-13}, Y_3 = L_{12} + L_{34} \end{cases} \tag{4.0}$$

The two other isometry Lie algebras will be described below.

As it follows from (1.1), (1.2), (1.3), (3.1), (3.4), and (3.5), the centers of d, f, and of l are of dimension one, and they are generated by L_{-10}, by L_{34}, and by $l_1 = (1/2) (L_{-13} + L_{04} - L_{-10} - L_{34})$, respectively.

Theorem 4.1: The following is true for the Lie algebra iso(D):

$$iso(D) = su(2) \oplus \{R\} \oplus su(2) = Z_{su(2,2)}(L_{-10})$$

(4.1)

Theorem 4.2: The following is true for the Lie algebra iso(F):

$$iso(F) = su(1,1) \oplus \{R\} \oplus su(1,1) = Z_{su(2,2)}(I_{34}).$$

(4.2)

Theorem 4.3: The centralizer of I_1 is a nine-dimensional Lie subalgebra of su(2, 2). More specifically, it is a two-dimensional nilpotent extension of iso(L).

Proof of the theorem 4.1: It is a straightforward exercise, based on (3.1), to verify that the second equality in the above (4.1) holds. Regarding the first equality, it immediately follows from (4.0) that one of the two su(2) blocks is generated by $\{X_1, X_2, X_3\}$, and the other is generated by $\{Y_1, Y_2, Y_3\}$ The center $\{R\}$ in (4.1) is generated by $X_0 = L_{-10}$ from (4.0).

Rather than to present other details from [3], one can use results from [16] to show that the basis (4.0) determines the Lie algebra iso(D). These results are based on the notion of a symmetric quadruple (k, q, p, B), where k is a finite dimensional real Lie algebra, q is a subalgebra of k, p is a non-zero, q-invariant, complementary vector subspace to q in k. B is a non-degenerate, q-invariant symmetric bilinear form on p. Also, [p, p] = q, and q contains no nonzero ideals of k. A certain involutive automorphism h of k is instrumental since q is its $\lambda = 1$ eigenspace, whereas p is its $\lambda = -1$ eigenspace.

It is shown in [16] that a simply connected pseudo-Riemannian symmetric space determines (up to an isomorphism) a symmetric quadruple. Given a symmetric quadruple, there exists a corresponding simply connected pseudo-Riemannian symmetric space. In our case iso(D) = k is reductive. To finish this alternative proof of equality (4.1), it is enough to present the corresponding symmetric quadruple (a Lorentzian one, due to the signature (+,-,-,-) of the form B on p). To do so, introduce the following linear bijection h of the Lie algebra k: h is an identical transformation on the subalgebra q spanned by $\{X_1 - Y_1, X_2 - Y_2, X_3 - X_3\}$, and h is negative 1 on the four-dimensional vector space p generated by

$$\{X_0, X_1 + Y_1, X_2 + Y_2, X_3 + Y_3\}.$$

Clearly, q is isomorphic to su(2).

It is a straightforward exercise to verify that h is an involutive isomorphism of k. The form B on p is introduced as a "pull-back" of the following invariant Lorentzian form on d=u(2): vectors X_0, X_1, X_2, X_3 are orthonormal, their scalar squares being 1, −1, −1, −1, respectively. This finishes the proof of theorem 4.1.

Let us now apply that last approach in the proof of theorem 4.2. The center R in (4.2) is generated by L_{34}. One of the su(1, 1)-blocks is generated by H_1, H_2, H_3 (these matrices have been introduced by (3.2) of Section 3); the other su(1, 1)-block in (4.2) is a subalgebra of su(2, 2) generated by

$$H_4 = L_{-10} + L_{12}, H_5 = L_{01} - L_{-12}, H_6 = L + L_{-11}.$$

Let us show that the seven-dimensional subalgebra h spanned by

$$H_0, H_1, H_2, H_3, H_4, H_5, H_6, \tag{4.3}$$

is the Lie algebra of the (reductive) Lie group Iso(F). The latter group is a subgroup of SU(2, 2). Introduce the following linear bijection t of the Lie algebra h: t is an identity transformation on the subalgebra q generated by

$$H_0 - H_4, H_1 - H_4, H_2 - H_6$$

(Clearly, q is isomorphic to su(1, 1)), and t is negative 1 on the four-dimensional vector space p generated by

$$H_0 + H_4, H_1 + H_5, H_2 + H_6, H_3 = L_{34}.$$

It is a straightforward exercise to verify that t is an involutive isomorphism of h. The form B on p is introduced as a "pull-back" of the following invariant Lorentzian form on f=u(1,1): vectors H_3, H_0, H_1, H_2 are orthonormal, their scalar squares being 1, −1, −1, −1, respectively. The corresponding Lorentzian quadruple is now (h, q, p, B).

Again, it is a straightforward exercise, based on (3.1), to verify that (4.2) holds, and that the above mentioned identity component of Iso(F) is the one of the block-diagonal subgroup of SU(2,2)(see [9, Theorem 9]). Theorem 4.2 is proven.

Proof of Theorem 4.3: It is known that the Lie algebra iso(L) is solvable and is of dimension 7. As a homogeneous symmetric Lorentzian manifold, L has been studied in [6]. The first author got to know the oscillator Lie group L from [17]. The L's important property to admit a non-degenerate bi-invariant metric has only been noticed in early 80s: [1, 2, and 18].

In paragraph 3 of [16] solvable Lorentzian quadruples are described in detail. That description includes a vector space w which is Euclidean in our case (the metric on w is chosen as a negative definite one). Also, a symmetric bilinear form A on w is part of the description. It is stated in [16] that if (k, q, p, B, w, A) is a solvable quadruple associated with a Lorentzian symmetric space L, then the full isometry Lie algebra iso(L) is the canonical semi-direct product of k with the algebra of skew-symmetric linear maps of w which commute with A. Here A is a linear operator associated with the form A. The operator A is defined by the formula

$$\mathbf{B}(Aw_1, w_2) = -\mathbf{A}(w_1, w_2).$$

The approach of [6] was based on a choice of four left-invariant vector fields on the oscillator Lie group L (their commutation relations are (1.1) from the Introduction). In (1.1), the vector field l_1 (which is also e_1

of (4.4) below) is both left-, and right-invariant. There are three more linearly independent right-invariant vector fields (they commute with left-invariant vector fields on L). Overall, in the approach of [6], we get the following table

$$
\begin{cases}
[e_2,e_3]=e_1,[e_2,e_4]=e_3, \\
[e_3,e_4]=-e_2,[e_5,e_6]=e_1, \\
[e_7,e_6]=e_6,[e_6,e_7]=e_5.
\end{cases}
$$

(4.4)

It determines the structure of the full isometry Lie algebra of the space-time in question. The seven generators, as elements of su(2, 2), can be chosen as follows (notice that this choice is different from the one in our Section 3):

$$e_1 = -\left(L_{-10} + L_{04} + L_{-11} + L_{14}\right),$$

$$e_2 = (1/2)\left(L_{-12} + L_{24} + 2L_{30} + L_{31}\right),$$

$$e_3 = (1/2)\left(L_{-13} + L_{34} + 2L_{02} + 2L_{12}\right),$$

$$e_4 = (1/8)\left(-5L_{-10} - 3L_{-11} + 3L_{04} + 5L_{14} + 4L_{23}\right),$$

$$e_5 = (1/2)\left(L_{-12} + L_{24} + 2L_{03} + 2L_{13}\right),$$

$$e_6 = (1/2)\left(L_{-13} + L_{34} - 2L_{02} - 2L_{12}\right),$$

$$e_7 = (1/8)\left(-5L_{-10} - 3L_{-11} + 3L_{04} + 5L_{14} - 4L_{23}\right).$$

Starting with (4.4), introduce w with an orthonormal basis e_5-e_3, e_2+e_6. Relative to this basis, the negative definite form $\langle .,. \rangle$, to be of use below, is given by the diagonal matrix

$$
\begin{bmatrix}
-1/2 & 0 \\
0 & -1/2
\end{bmatrix}.
$$

Introduce w^* with a reciprocal basis e_5+e_3, e_6-e_2. This involves a choice of a linear bijection between w and w^*, with w^* being the image of w, etc. Choose the form A given by the matrix

$$\begin{bmatrix} 2 & 0 \\ 0 & 2 \end{bmatrix}$$

relative to the same basis in w. Let us denote by $R, \overline{R}$, one-dimensional vector spaces generated by e_1, $e_4 + e_7$, respectively.

One can now verify that a six-dimensional

$$k = w^* \oplus R \oplus w \oplus \overline{R}$$

satisfies the following commutation table:

$$[k, R] = 0, [w^*, w^*] = [w, w] = 0, [e_7 + e_4, w] = w^*,$$

for w in w;

$$[\lambda, w] = \mathbf{A}(\lambda^*, w) e_1,$$

for λ in w^*, λ^* in w;

$$[\lambda, e_7 + e_4] = A(\lambda^*),$$

For λ^* in w.

Now, $q = w^*$, $P = \overline{R} \oplus w \oplus R$, and B is defined as follows: on w, B coincides with $\langle .,. \rangle$, from above,

$$\mathbf{B}\left(\overline{R}\oplus R,w\right)=\mathbf{B}\left(\overline{R},\overline{R}\right)=\mathbf{B}\left(R,R\right)=0\,,$$

$$\mathbf{B}\left(e_1,e_4+e_7\right)=1.$$

This is a particular case of commutation relations from p.588 of [2]. Vector e_7-e_4 generates the algebra of those skew-symmetric linear maps of w which commute with A. We have thus applied results of [16] to show that the above seven-dimensional Lie algebra is the entire isometry Lie algebra of the corresponding symmetric Lorentzian space. To finish the proof of our theorem 4.3, we provide two more generators, L_{23}, $L_{-11}+L_{14}$ which (together with the above vectors $e_1,\ldots,e_7$) form the centralizer of e_1. One can check the commutation relations to show that this nine-dimensional Lie subalgebra of su(2,2) is a nilpotent extension of the seven-dimensional iso(L).

Remark 4.1: Presumably, if one adds the generator of a homothetic transformation (which acts non-trivially on e_1), then the resulting 10-dimensional Lie algebra is the one discovered on p.130 of [19].

Remark 4.2: A centralizer of a non-zero element in su(2, 2) may have dimension 3, 5, 7, or 9. It seems to be of interest to try to characterize those cases when the isometry Lie algebra coincides with the centralizer of a single element from the Lie algebra of all conformal transformations: it is so in theorems 4.1, 4.2 but it is not the case of theorem 4.3. To continue, recall that k = iso(D) is defined by our (4.0).

Theorem 4.4: There exists an su(2, 2)-inner contraction of k to iso(L).

Proof: As in (3.3), conjugate the involved matrices by a(s, t) and set t = 0. Then choose the following seven vectors in the resulting Lie algebra aka^{-1}:

$$L_1=-a\left(s,0\right)\left(L_{-10}+L_{34}-L_{12}\right)a^{-1}\left(s,0\right),$$

$$L_2=a\left(s,0\right)\left(L_{13}+L_{24}\right)a^{-1}\left(s,0\right),$$

$$L_3 = a(s,0)(L_{14} - L_{23})a^{-1}(s,0),$$

$$L_4 = -(1/2)a(s,0)(L_{-10} - L_{34} + L_{12})a^{-1}(s,0),$$

$$L_5 = a(s,0)(L_{24} - L_{13})a^{-1}(s,0),$$

$$L_6 = a(s,0)(L_{14} + L_{23})a^{-1}(s,0),$$

$$L_7 = (1/2)a(s,0)(L_{-10} + L_{34} + L_{12})a^{-1}(s,0).$$

In the limit as s goes to infinity, the seven matrices form the following Lie subalgebra, isomorphic to iso(L):

$$l_1 = -(1/2)(L_{-10} + L_{34} - L_{-13} - L_{04}),$$

$$l_2 = (1/2)(L_{13} + L_{01} + L_{24} - L_{-12}),$$

$$l_3 = (1/2)(L_{14} - L_{-11} - L_{23} - L_{02}),$$

$$l_4 = -(1/2)(L_{-10} - L_{34} + L_{12}),$$

$$l_5 = (1/2)(-L_{13} - L_{01} + L_{24} - L_{-12}),$$

$$l_6 = (1/2)(L_{14} - L_{-11} + L_{23} + L_{02}),$$

$$l_7 = (1/2)(L_{-10} - L_{34} - L_{12}).$$

$$(4.5)$$

Remark 4.3: It is obvious that the Lie algebra (3.8) is a subalgebra of (4.5).

We can contract h=iso(F) in, essentially, the same way we did the contraction of k=iso(D) This h is $Z_{su(2,2)}(L_{34})$ which is generated by the following vectors:

$$L_{-10}, L_{-11}, L_{-12}, L_{01}, L_{02}, L_{12}, L_{34}.$$

Theorem 4.5: There exists an su(2, 2)-inner contraction of h = iso(F) to iso(L).

Proof: Complete basis (3.3) with vectors

$$H_4 = L_{01} - L_{-12}, H_5 = L_{02} + L_{-11}, H_6 = L_{-10} - L_{34} + L_{12}$$

to obtain a basis for iso(F), and extend the contraction, defined earlier for f, to this algebra. Namely, conjugate

$$H_0, H_1, H_2, H_3, H_4, H_5, H_6$$

by a(t, s) introduced in Section 3. Setting t = 0 and letting s go to infinity, we get a Lie algebra generated by the vectors

$$L_1 = (1/2)(L_{-10} + L_{34} - L_{-13} - L_{04}),$$

$$L_2 = (1/2)(L_{01} + L_{13} + L_{-12}),$$

$$L_3 = (1/2)(L_{02} + L_{23} - L_{-11} + L_{14}),$$

$$L_4 = -(1/2)(L_{-10} - L_{34} - L_{12}),$$

$$L_5 = (1/2)(L_{01} + L_{13} - L_{-12} + L_{24}),$$

$$L_6 = (1/2)(L_{02} + L_{23} + L_{-11} - L_{14}),$$

$$L_7 = -(1/2)(L_{-10} - L_{34} + L_{12}).$$

This last Lie algebra is (isomorphic to) iso(L).

ACKNOWLEDGMENTS

A.L. thanks David Vogan for helpful discussions on the subject of Section 3, Ernest Vinberg for information on centralizers in su(2, 2), and the comments for Referee.

REFERENCES

1. A. K. Guts and A. V. Levichev, "On the Foundations of Relativity Theory," Doklady Akademii Nauk SSSR, Vol. 277, No. 6, 1984, pp. 1299-1303. (in Russian)
2. A. V. Levichev, "Causal Cones in Low-Dimensional Lie Algebras," Siberian Journal of Mathematics, Vol. 26, No. 5, 1985, pp. 192-195. (in Russian)
3. S. Paneitz and I. Segal, "Analysis in Space-Time Bundles I: General Considerations and the Scalar Bundle," Journal of Functional Analysis, Vol. 47, No. 1, 1982, pp. 78-142.http://dx.doi.org/10.1016/0022-1236(82)90101-X
4. A. V. Levichev, "Certain Symmetric General Relativistic Space-Times as the Solutions to the Einstein-Yang-Mills Equations,", Proceedings Group Theoretical Methods in Physics (III International Seminar), Yurmala, 1985, pp. 145-150. (in Russian)
5. [5]D. Kramer, H. Stephani, M. MacCallum and E. Herlt, "Exact Solutions of Einstein's Field Equations," VEB Deutscher Verlag der Wissenschaften, Berlin, 1980.
6. A. V. Levichev, "Chronogeometry of an Electromagnetic Wave Defined by a Bi-Invariant Metric on the Oscillator Lie Group," Siberian Journal of Mathematics, Vol. 27, No. 2, 1986, pp. 237-245. http://dx.doi.org/10.1007/BF00969391
7. JHilgert, K. H. Hofmann and J. D. Lawson, "Lie Groups, Convex Cones, and Semigroups," Clarendon Press, Oxford, 1989.
8. A. V. Levichev, "Three Symmetric Worlds Instead of the Minkowski Space-Time," Transactions on RANS, series MMM&C, Vol. 7, No. 3-4, 2003, pp. 87-93.
9. A. V. Levichev, "Pseudo-Hermitian Realization of the Minkowski World through the DLF-Theory," Physica Scripta, Vol. 83, No. 1, 2011, pp. 1-9.
10. V. Guillemin and S. Sternberg, "Geometric Asymptotics," American Mathematical Society, Providence, 1977. http://dx.doi.org/10.1090/surv/014
11. A. Fialowski and M. De Montigny, "On Deformations and Contractions of Lie Algebras," SIGMA, Vol. 2, 2006, p. 10. http://www.emis.de/journals/SIGMA/2006/Paper048/
12. I. Segal, "A Class of Operator Algebras Which Are Determined by Groups," Duke Mathematical Journal, Vol. 18, No. 1, 1951, pp. 221-265. http://dx.doi.org/10.1215/S0012-7094-51-01817-0
13. A. Knapp, "Representation Theory of Semisimple Groups: An Overview Based on Examples," Princeton University Press, Princeton, 2001.
14. I. E. Segal, H. P. Jakobsen, B. Orsted, S. M. Paneitz and B. Speh, "Covariant Chronogeometry and Extreme Distances: Elementary Particles," Proceedings of the National Academy of Sciences, Vol. 78, No. 9, 1981, pp. 5261-5265.http://dx.doi.org/10.1073/pnas.78.9.5261
15. S. Sternberg, "Chronogeometry and Symplectic Geometry," Colloques Internationaux C.N.R.S. Geometrie Symplectique et Physique Mathematique, Vol. 237, 1975, pp. 45-57.

16. M. Cahen, and N. Wallach, "Lorentzian Symmetric Spaces," Bulletin of the American Mathematical Society, Vol. 76, No. 3, 1970, pp. 585-591. http://dx.doi.org/10.1090/S0002-9904-1970-12448-X
17. R. F. Streater, "The Representations of the Oscillator Group," Communications in Mathematical Physics, Vol. 4, No. 3, 1967, pp. 217-236. http://dx.doi.org/10.1007/BF01645431
18. A. Medina and Ph. Revoy, "Les Groups Oscillateurs at Leurs Reseaux," Manuscripta Mathematica, Vol. 52, No. 1-3, 1985, pp. 81-95. http://dx.doi.org/10.1007/BF01171487
19. M. Cahen and Y. Kerbrat, "Champs des Vecteurs Conformes et Transformations Conformes des Espace Lorentziens Symmetriques," Journal de Mathématiques Pures et Appliquées, Vol. 4, No. 57, 1978, pp. 99-132.

CITATION

1. A. Levichev and O. Sviderskiy, "Contractions of Certain Lie Algebras in the Context of the DLF-Theory," Advances in Pure Mathematics, Vol. 4 No. 1, 2014, pp. 1-10. doi: 10.4236/apm.2014.41001.

Index

A

algebraic geometry 11

C

cellular phone 93, 94, 96
Credit card number 93, 95, 106, 107, 108

D

Discrete topology 237, 241

H

homeomorphisms 6, 23
homomorphism 1, 49, 52, 64, 65, 67, 75, 78, 79, 80, 85
homotopy 1, 2, 3, 5, 6, 7, 8, 10, 11, 12, 13, 14, 16, 17, 18, 19, 20, 21, 22, 23, 24, 25, 26, 27, 30, 36, 40, 41, 46, 47, 48, 49, 50, 51, 52, 53, 54, 55, 58, 59, 60, 61, 67, 68, 69, 70, 73, 74, 76, 77, 79, 81, 82, 83, 85, 86, 87, 88, 89, 90, 91
homotopy equivalences. 2
homotopy groups 1, 2, 6, 8, 10, 11, 13, 14, 16, 17, 30, 36, 47, 49, 52, 59, 61, 67, 68, 73, 79, 85, 89

M

Memory capacity 94

Monty Hall problem 238

P

Philosophy 236
Posterior state 236, 240
Probability 93, 94, 95, 96, 98, 99, 100, 101, 102, 105, 107

Q

Quantum mechanics 227, 228, 230, 232, 233, 235

S

Spectrum space 231
sphere spectrum 4, 17
Steenrod algebra 1, 2, 10, 38, 62
Stochastic variable 98
Storage capacity 93
suspension spectrum 4, 18, 19, 31

T

Thom spectrum 8, 27, 30
topological cyclic 7, 11, 17, 18, 23, 24, 88, 89
topological cyclic homology 7, 11, 17, 18, 23, 24, 88, 89